The Companion Guide to Teaching Maths

By

Bryan Whelan

Introduction

The purpose of this book is to provide alternative methods for teaching some of the basic Mathematical concepts at a Secondary level. It is a guide and a reference for Maths teachers who may be struggling or perhaps, disillusioned with the accepted explanations in the Standard Maths Textbooks. It is also intended to stimulate and provide alternative approaches to Mathematical concepts.

A standard text book usually provides an explanation of how to perform a process in Maths. However, not all students will 'get it'. This guide is designed to provide you, the teacher, with another way of doing it. Hopefully, giving the student an opportunity to be successful. Also the methods described in this book are aimed at reducing the amount of facts and processes the student needs to remember to help master the various concepts more easily.

I have purposely not included exercises in the book. The traditional Maths text book has a plethora of exercises for the student. These days, there are also many useful exercise sheets available on the internet.

The topics I have covered, are aimed at Years 7 – 10. I purposely did not section off aspects of the topics into year levels. Students tend to learn Maths at different rates. It is just as likely that a Year 7 student could happily handle surds say, and a Year 10 student struggle with basic integers. It is left to you, the teacher, to gauge how far into the topic you want to go for each of your students.

Similarly I have not covered every single Maths topic you may encounter in Years 7 – 10. Sometimes the topic really only has one way of explaining the concept. However, further topics may be looked at for another Companion Guide.

About the Author:

Bryan Whelan was a Maths teacher for over 30 years. During this time, he developed a lot of the methods shown in this book as a result of seeing many students struggling with the normal explanation in the standard text books. He taught Maths from Year 7 to year 12 at several different schools, teaching both old and new curricula.

Contents:

Multiplication

When students embark on Secondary College Mathematics, they should really know their basic times tables (or at least be able to 'skip count', 3,6,9,12 etc.).

Unfortunately not all do. So it falls upon the Maths teacher to make sure they do. The following exercises are some ideas to help train students in remembering basic times tables.

The 'Short' Times Table

Back when I went to school, I was taught to memorise times table up to 12 x 12. But what are the times tables really used for? They provide the basis for all multiplication and division. If you look at those processes closely, you will find that students only really need to know times tables up to 10. I.e. from 1 x 1 up to 10 x 10.

Also, each times problem has the same answer not matter which order they are in.

E.g. 5 x 6 = 30, 6 x 5 = 30

So the actual number of times problems to memorise is reduced by half (in fact if you wanted to count them, there are 55 multiplications to memorise in all). If we further discount some of the easiest ones, like the 1 x and the 10 x (which you would hope, everyone knows), this brings the number down to just 36 multiplications.

Here they are:

2 x 2 = 4 2 x 3 = 6 2 x 4 = 8   2 x 5 = 10 2 x 6 = 12 2 x 7 = 14 2 x 8 = 16 2 x 9 = 18

3 x 3 = 9 3 x 4 = 12 3 x 5 = 15 3 x 6 = 18 3 x 7 = 21 3 x 8 = 24 3 x 9 = 27

4 x 4 = 16 4 x 5 = 20 4 x 6 = 24 4 x 7 = 28 4 x 8 = 32 4 x 9 = 36

5 x 5 = 25 5 x 6 = 30 5 x 7 = 35 5 x 8 = 40 5 x 9 = 45

6 x 6 = 36 6 x 7 = 42 6 x 8 = 48 6 x 9 = 54

7 x 7 = 49 7 x 8 = 56 7 x 9 = 63

8 x 8 = 64 8 x 9 = 72

9 x 9 = 81

Some students will further know some of them in this list already. So they need only focus on the ones they can't remember.

Simple tricks like writing them down each lesson in their books as part of a routine, will help them. Similarly, there is a plethora of on line games that they could play to reinforce their knowledge of these multiplications.

Patterns

If you ask your students to remember a number for next lesson, the chance that they will, may probably depend on what that number is? For instance, to recall (in say a week's time) the number 376294 would be a much more difficult task than say the number 123456. Why? Simple! The second number has a set pattern to it. You really only need remember the first number and the pattern.

This idea can be transferred to helping students remember their multiplication tables. For instance:

2 times table – all answers end in even numbers. (all answers end in 0, 2, 4, 6, 8) . Same for 4, 6, 8 times tables.

3 times table – double the number then add it again

E.g. 3 x 8 – double 8 = 16 add another 8 16 + 8 = 24 So 3 x 8 = 24

4 times table – double the number then double again

E.g. 4 x 8 – double 8 = 16 double 16 = 32 So 4 x 8 = 32

5 x table – all answers end in either 5 (if the number is odd) or 0 (if the number is even).

E.g. 5 x 3 = 15 (3 is odd); 5 x 6 = 30 (6 is even)

Also, to calculate, say, 5 x 8, add a 0 to the 8 (80) and then halve it. Answer 40.

6 times table- Again add a 0 to the number, halve the result then add the original number.

E.g. 6 x 4 = ? Add a 0 to 4 – 40, halve it – 20, then add the original 4 – 24 6 x 4 = 24

NOTE: whenever you multiply 6 by an even number the last digit is the same

E.g. 6 x 2 = 12 6 x 4 = 24 6 x 6 = 36 6 x 8 = 48 6 x 10 = 60

7 times table – No real easy one for these but you might try adding a 0 then subtracting 3 times the original number

E.g. 7 x 3 - add 0 to 3 - 30 3 times 3 = 9, so subtract 9 – 30 -9 = 21 So 7 x 3 = 21

8 times table – Similarly for these ones, you could add a 0 and subtract 2 times the original number

E.g. 8 x 4 - add 0 to 4 - 40 2 times 4 = 8, so subtract 8 – 40 -8 = 32 So 8 x 4 = 32

9 times table- There are a few method for these ones.

Method 1 - add a 0 then subtract the original number

E.g. 9 x 4 - add 0 to 4 - 40 subtract 4 – 40 -4 = 36 So 9 x 4 = 36

Method 2 – subtract 1 from the original number. That is your first digit. The other digit is that one subtracted from 9.

E.g. 9 x 4 - 4 – 1 = 3 3 + 6 = 9 So 9 x 4 = 36

Method 3 – Use your fingers. Hold your fingers out in front of you. Imagine they are number from the left 1 – 10. To multiply 9 x 4, say, bend finger number 4. The answer is then the number of fingers

There are many other patterns you might find out about, but with reinforcement, students eventually get used to multiplication tables

Calculator or no calculator

In the early days of calculators, most Maths teachers refused to have them in the classroom, claiming that they made students lazy thinkers. While that is possibly still true today, we all use calculators for one thing or another.

So, should they be used in class? If you analyse your own use of calculators, you will find the answer. That is, you would use the calculator for the 'more complex' calculations. So for students, the advice is to only allow calculators for those problems they shouldn't be able to work out in their head.

For instance, students should not be using calculators for doing basic times tables. They should use them for say, long division. So when a student should not be using a calculator will vary between students.

All students, however, should be aiming to be able to multiply a single digit number by any two digit number in their head. Also they should be able to do most simple divisions and addition and subtractions for double digits. So ideally, these processes should not be done on a calculator.

How can this be done?

1. An experiment to convince students when to use calculators. This is a speed test to see who can get the correct answer first. Select 2 students from the class, one to use a calculator and the other to use pen and paper or work it out in their head. The student using the calculator must use the calculator and nothing else. The other student can use anything but NOT the calculator.

   The first calculation to try would be something like 23 x 56 = 1288. The second could be say, 12 x 65 = 780. The student using the calculator will win both times. Now for the third one, we apply our trick and use a problem like 2 x 4 = 8. The student using their head should know the answer straight off and win. The student using the calculator MUST press the buttons to get the answer. This experiment demonstrates that thinking with your brain is much quicker. So aiming to be able to do as many calculations as possible in your head is much more efficient.

2. Have a set lesson where no calculators are used. You may have to be careful to select your problems accordingly for different students but it forces them to try and calculate in their head.

3. Have a weekly game session on Arithmetic. A good one to try is the Letters and Numbers game (see below). Or perhaps a mini test at the start of each lesson on some aspect of Arithmetic.

Extension Activities

Letters and Numbers

This was a TV show a few years back which used letters for making words and numbers for calculating totals. For the numbers part, the contestant chose 6 numbers. Of the 6, some could be 'small' (which were random numbers between 1 and 10) or large (the numbers 25, 50, 75 or 100). A random total was then selected and the contestants had to combine some or all of the numbers using add, subtract, multiply or divide (and even brackets) to get to the total (or as close as possible). You

can only use each number once. This could be used as a game in class (or even as homework or extension work)

Here are some samples you can try:

1. 9, 3, 1, 50, 75, 100 The target number is 479
2. 10, 4, 1, 5, 5, 9 The target number is 911
3. 2, 8, 7, 8, 25, 75 The target number is 643

Solutions

1. 9 x 50 = 450, 450 + 100 – 75 = 475, 475 + 1 + 3 = 479
2. 5 x 5 = 25, 25 x 4 = 100, 100 x 9 = 900, 900 +10 + 1 = 911
3. 8 – 2 = 6, 6 x 75 = 450, 450 + (8 x 25) = 650, 650 – 7 = 643

Some Calculator Puzzles

There are several neat tricks you can use with a calculator. Here are a few:

1. Start with any 5 digit number. Now re arrange the same 5 digits to form another number. Subtract the smaller one form the larger one. Now add up the digits in the answer. Repeat this until you have a single digit. The answer will always be 9! Example. Start with 42178. Rearrange the digits – 18724. Subtract 42178 – 18724 = 23454. Add up the digits – 2+3+4+5+4 = 18. Repeat. 1 + 8 = 9!

2. Game of Nim. Two students can play against each other. Using a calculator, start with the number 21. Each player in turn can subtract either 1, 2 or 3. The loser is the one who ends up with 0. A variation of this is to have students subtract any digit in the first column of their calculator (1, 4 or 7) instead of 1, 2 or 3.

3. Number patterns. Try the following:

   a. $11^2 =$ $111^2 =$ $1111^2 =$ etc. There should be a pattern emerge
   b. 101 x 1 = ? 101 x 101 = ? 101 x 101 x 101 = ? etc. Again you should find a pattern.
   c. Dividing any whole number by 11 gives some pretty interesting decmials.

Napier's Bones

John Napier was a Scottish Mathematician who invented Logarithms. To demonstrate the process he constructed a set of rods (made of whalebone – hence the name), which when put together, let you find the answer to any multiplication problem by a single digit.

You can construct your own Napier's Bones out of thin cardboard. Copy the following diagram below onto card and cut them into a series of vertical rods. You may already notice the pattern of numbers on the rods. The 'M' rod is the Multiplying rod. Once done, you should have 10 rods each numbered 0 to 9 at the top and the Multiplying rod.

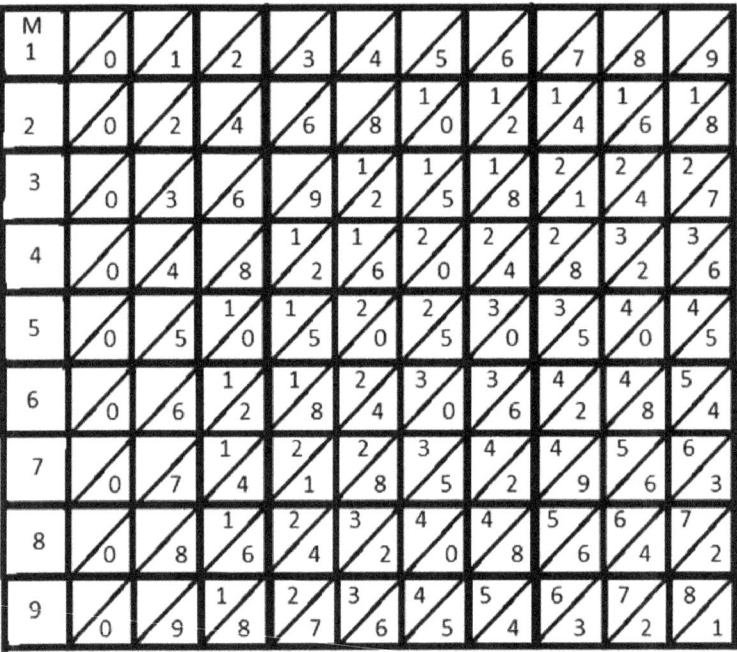

How to Use:

Examples

1. 23 x 3   Put the '2' rod next to the '3' to make 23. The put the Multiplying rod next to them thus:

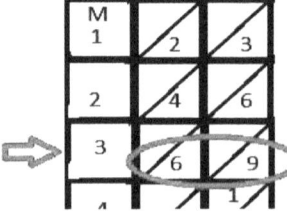

The answer is 69

2. 72 x 9   Put the '7' rod next to the '2' and the Multiplying rod next to them, thus:

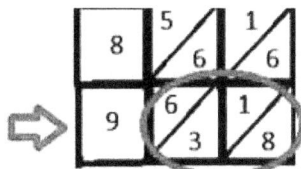

When you have 2 numbers in the middle, i.e. the '3' and the '1', you add them.

So the answer is 648

3. 69 x 8   Again put the '6' rod next to the '9' rod and the Multiplying rod next to them, thus:

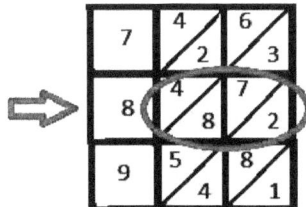

Again add the middle numbers, 8 + 7 = 15.

In this case, you 'carry the 1' to the first digit.

So the answer is 552

Multiplying single digit by double digits in your head!

Yes it can be done! Try it with the following examples:

E.g. 6 x 43 (In your head, do 6 x 3 = 18, remember the 1 and 8. Now do 6 x 24 = 24. Add the 1 from before to get 25. Now put the 8 on the end. Answer 258)

1. 7 x 38 (7 x 8 = 56, remember the 5 and 6. 7 x 3 = 21. Add 5 to get 26. Put the 6 on the end. Answer 266)
2. 3 x 72 = 216
3. 7 x 68 = 476

This can be done as a class or individual exercise. Or as part of your lesson where you don't use a calculator. You can encourage students to spend five minutes or so at home, trying this in their head. The frequency of repetition is key to them getting used to doing this. If they have trouble with doing it, get them to try the multiply forwards method mentioned later. It is a little easier.

Squaring any 2 digit number in your head!

A good party trick. Using the trick from above, this is how it can be done. Let's try with 482

Step 1: Add to (or subtract from) the number to get to the next multiple of 10 and take note of how much you added (or subtracted). Then subtract (or add) the same amount from the original number. This will give you two numbers.

For our example, 48, nearest multiple of 10 is 50 (add 2). So subtract 2 as well to give you the numbers 50 and 46.

Step 2: Multiply these two numbers in your head. Keep in mind that with the multiple of 10, you need only multiply by the single digit (as above).

For our example, we actually only multiply 5 x 46 = 230.

Step 3: Square the number you added (or subtracted) and add that number to the end. You have your answer

For our example, we originally added 2. 2^2 = 4. So our final answer is 2304.

Here are a few more examples to show you how it works:

Second example: 83^2

Nearest multiple of 10 is 80. So subtract 3 to get 80 and add 3 to get 86.

So 8 x 86 = 688

Squaring 3 we get 9. So our final answer is 6889

Third example: 74^2

Nearest multiple of 10 is 70. So subtract 4 to get 70 and add 4 to get 78.

So 7 x 78 = 546

Squaring 4 we get 16. (This is our only complication as you now have to add 16 to the end. You add the 6 and carry the 1). So our final answer is 5476

A faster way to do long multiplication

E.g.   27
         38 x

Step 1: Multiply the two end digits 7 x 8 = 56 . Put down the 6 and carry the 5
   5   27
        38 x
         6
Step 2: 'Cross Multiply' and add. That is for our example 8 x 2 = 16 and 3 x 7 = 21. 21 + 16 = 37. Also add anything carried over. I.e. 37 + 5 = 42. Again put down the 2 and carry the 4

4   27
       38 x
       26

Step 3: Now multiply the first two digit and add anything carried over. For our example, 2 x 3 = 6 + 4 = 10. So our final answer is

      27
      38 x
      1026

A second example:

      32   Multiply the 3 end numbers. 7 x 2 = 14 (put down the 4, carry the one)
      27 x

1   32   cross multiply: 3 x 7 = 21 and 2 x 2 = 4, 21 + 4 = 25 and add the carried 1,
      27 x   25+1 =26 (put down the 6 and carry the 2)
       4

2   32   Multiply the two front numbers. 3 x 2 = 6 and add the carried over 2, 6 + 2 = 8
      27 x   (put down the 8)
      64

      32   Final Answer
      27 x
      864

Here are a few other examples to try:

E.g. 1. 26 x 53 = 1378
      2. 47 x 72 = 3384
      3. 54 x 81 = 4374

A new way of multiplying large numbers (in your head!)

Method 1: Multiply forwards

E.g. 4 x 237 = ?
     (Normally you would multiply this by starting at the right and move left. However, we read the number left to right. So why not multiply left to right!)

Step 1: Start at the left and multiply. 4 x 2 = 8. Put down the 8
4 x 237 = 8
Step 2: Repeat this for each subsequent digit. Our second one is 4 x 3 = 12. Like in normal multiply we put down the 2 and carry the 1. This gives us:
4 x 237 = 92
Now repeat for the last digit: 4 x 7 = 28 Put down the 8 and carry the 2
4 x 237 = 948

     Note: The difficult part of this is the carry part. We can overcome this somewhat by using a 'look ahead' check.

Returning to the above problem, when we multiply we can 'look ahead' to see if there is anything to carry across.

I.e. 4 x 237 = ? We firstly do 4 x 2 but we look ahead and check the next one 4 x 3 = 12. So our 4 x 2 = 8 and we add the 1 immediately.

4 x 237 = 9

Next we repeat for the next digit. 4 X 3 = 12 (we only need the 2). Look Ahead to the next one 4 X 7 = 28. So add 2. 4 X 3 = 12 add the 2 gives us 4

4 x 237 = 94

Finally we do the last digit. 4 x 7 = 28 (only need the 8)

4 x 237 = 948

Here are some other examples to try: (You can use this method no matter how big the numbers)

1. 3 x 385 = ?
2. 4 x 1234 = ?
3. 6 x 35712 = ?

Method 2 – Use base numbers

Let's start with a base number of 10

E.g. 13 x 12 (answer is 166)

Step 1: Find out how much over 10 is each number

13 x 12

+3   +2

Step 2: Add one of the numbers to one of the 'over' numbers. (It doesn't matter which you choose as the answer will be the same). This will give the first part of your answer.

13 + 2 = 15 (NOTE: the 12 +3 also give 15)

Step 3: Now multiply these 2 'over' numbers.

3 x 2 = 6. This is the last part of your answer

13 x 12 = 156

Another example

18 x 14 = ?

Step 1:   +8   +4

Step 2:   18 + 4 = 22

Step 3:   8 x 4 = 32 (Put down the 6 and carry the 5)

I.e. you are really add 220 + 32 = 252

So 18 x 14 = 252

NOTE: This trick works for any numbers between 10 and 19. To do this for higher numbers, say numbers in the twenties, then use the base number of 20. Then when you do step 1, you have to multiply by 2. For thirties, multiply by 3 etc. Here is an example:

27 x 24 = ? Use base 20

Step 1: +7   +4

Step 2: 27 + 4 = 31 but as it is base 20, then double this - 62

Step 3: 7 x 4 = 28. Add to 620

Answer 648

Teaching Points

1. All of these mental methods require practice and concentration. It pays to start with easy problems first to give them some confidence. Also doing them often reinforces the process better.
2. Most of these methods work best by demonstration rather than by remembering steps.
3. Also there are several very good videos and websites on the internet. Just search for multiplication and division tricks.

Division and Fractions

Division

Consider a simple example: $12 \div 4$. The answer is 3 because $4 \times 3 = 12$. This illustrates the first basic principle of division that:

> ## **Divide is the opposite of multiply.**

A simple concept which most students should already understand but it is worth emphasising.

E.g. if $4 \times 5 = 20$ then $20 \div 4 = 5$ (also $20 \div 5 = 4$)

The second principle is not often emphasised but should be. Consider again $12 \div 4$. We could have expressed this as a fraction:

$\frac{12}{4}$ Cancelling down (see below) we get $\frac{3}{1} = 3$

The second principle of divide is:

> ## **A divide is a fraction and a fraction is a divide.**

The main advantage of this is simply that you can rewrite any divide as a fraction and simplify or rewrite any fraction as a divide.

E.g. $128 \div 16 = ?$ Rewrite as fraction and simplify

$\frac{128}{16}$ = (both divide by 4 say) = $\frac{32}{4}$ = (and divide by 4 again) = $\frac{8}{1} = 8$

Where it can be very handy, is with divides that don't give a nice answer. Try this one on your students:

$1 \div 3 = ?$ Students find this hard for some reason. But rewrite it as a fraction and you have your answer — $^1/_3$!

E.g.

1. $2 \div 5 = ?$ Answer $^2/_5$
2. $125 \div 55 = ?$ Answer $= {}^{125}/_{55} = {}^{25}/_{11} = 2^3/_{11}$ (if you want the answer as a mixed number. See later)

Teaching Points:

1. In some cases the answers required need be written as decimals or perhaps mixed numbers where appropriate. Try to leave these conversions until the students are fully comfortable with the fraction answer. Also this will destroy the myth that decimal and mixed number answers are more correct.

2. Like all arithmetic problems, it is a good idea to reinforce the notion of estimating the answer in your before you start. This simply means to pick 'nice' numbers that are close and estimate the answer.

   E.g. $125 \div 55 = ?$ (from above). Pick numbers that are close but easy to work with. For instance 125 is near 100 and 55 is near 50. So $100 \div 50 = 2$. I.e. the answer is going to be something near 2. This technique is also useful if you use a calculator

3. Most students will use a calculator for divide problems which is fine. However, they should be encouraged to avoid the calculator for the more simple problems and make use of the estimation trick above to help get the right answer.

Extension Work:

Tricks for division

1. I hope the obvious one you should have tumbled to by now is convert the divide into a fraction and simplify.

   E.g. 420 ÷ 35. Convert to $\frac{420}{35}$. Both divide through by 5. This gives $\frac{420 \div 5}{35 \div 5} = \frac{84}{7}$

   (You could convert back at this point if you like.) 84 ÷ 7 = 12

2. Split the dividing number (the divisor) into the product of two other numbers that you are more familiar with.

   E.g. 1596 ÷ 42. 42 is 6 x 7. So divide by 6 first. This gives 1596 ÷ 6 = 266. Now divide by 7. This gives 266 ÷ 7 = 38. So 1596 ÷ 42 = 38

Fractions

For the average student, the most important parts of learning fractions, falls into 3 areas. Knowing physically how much $2/3$ actually looks like; knowing that ¾ is larger than $2/5$ and the ability to be able to cancel down. These has more practical value than being able to add or subtract two fractions.

How big are these fractions?

Trick 1: Check the top number of the fraction. Double it. If the answer is bigger than the bottom number them the fraction is more than ½, which indicates that the fraction is large.
E.g.
1. $12/19$ Double 12 = 24. This is more than 19. So the fraction is bigger than ½ (large)
2. $2/5$ Double 2 = 4 This is less than 5 So the fraction is less than ½ (small)

Trick 2: How close are the top and bottom numbers? The closer they are the closer the fraction is to 1.
E.g.
1. $15/17$ 15 is close to 17, so the fraction is close to 1.
2. $4/18$ 4 is not close to 18, so the fraction is closer to 0

Which fraction is bigger?

Trick 1: Compare two fractions - Which fraction is bigger? Compare the top numbers and the bottom numbers. The bigger fraction will have both numbers bigger.

Examples:

1. Which is bigger, $3/7$ or $4/9$? Since 4 is bigger than 3 and 9 is bigger than 7, $4/9$ is bigger than $3/7$

2. Which is bigger, $^7/_8$ or $^5/_6$? Since 7 is bigger than 5 and 8 is bigger than 6, $^7/_8$ is bigger than $^5/_6$

Trick 2: From Trick 1 above, what happens if one of the numbers is bigger but not the other. The trick is to multiply each top number by the other bottom number. The one with the bigger total is the bigger fraction.

Examples:

1. Which is bigger, $^3/_7$ or $^2/_9$? Multiply 3 x 9 = 27 and 2 x 7 = 14. Since 27 is bigger than 14 then $^3/_7$ is the bigger fraction.

2. Which is bigger, $^5/_8$ or $^6/_7$? Multiply 5 x 7 = 35 and 6 x 8 = 48. Since 48 is bigger than 35 then $^6/_7$ is the bigger fraction.

Trick 3: The Fraction Chart. This is a chart that starts with a block. It is divided into ½ then $^1/3$ etc. Looks something like this:

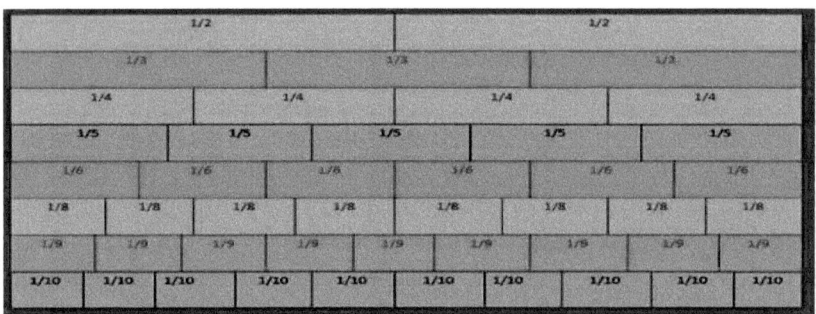

From the chart, you can see that a fraction like $^1/_{10}$ is smaller than $^1/_4$. Also $^2/_9$ is bigger than $^1/_5$.

Cancelling down and Equivalent Fractions

Equivalent fractions are those which have different top and bottom numbers but have the same value. For instance ¾ and $^6/_8$. The connection between the two or any equivalent fraction is simply that the top and bottom number have been **multiplied or divided** by the same number (In our example, ¾ has been multiplied top and bottom, by 2).

The equivalent fraction with the lowest pair of wholes numbers is called the fraction in lowest (or simplest) form. To find this, we divide top and bottom by the same number and repeat until we can't go any further. This is called **cancelling down**.

Example:

Cancel down $^{36}/_{80}$ to its lowest form. $^{36}/_{80}$ -Divide top and bottom exactly by 4 to give $^9/_{20}$. Since no number divides exactly into both 9 and 20, this fraction is now is lowest form.

There is no real shortcut to doing this but knowing some divisibility tricks can help.

Examples:

Cancel down each of the following:

1. <u>6</u>. Knowing that any even number divides by 2, you can divide each of these by 2   <u>6</u> = <u>3</u>
   38   38   19
2. <u>18.</u> Each of these numbers divide by 3 (see tricks below). So <u>18</u> = <u>6</u> = <u>2</u>
   27   27   9   3

Tricks for knowing what numbers are divisible by what:

Divisible by 2 : Number is even   (i.e. ends in 0,2,4,6, or 8) E.g. 284
Divisible by 3 : Digits add up to 0,3, 6 or 9 E.g. 234 (since 2+3+4 is 9)
Divisible by 4 : Last two digits are divisible by 4. E.g. 131740
Divisible by 5 : Numbers end in 0 or 5 E.g. 2535
Divisible by 6 : Digits add up to 0,3, 6 or 9 AND the number is even. E.g. 5316
Divisible by 7 : Double the last digit. Subtract that from the other digits. If the answer is divisible by 7 then the number is divisible by 7 E.g. 392 Double last digit = 4, 39 – 4 = 35, 7 x 5 = 35 So number is divisible by 7
Divisible by 8 : Take the last 3 digits. If you can halve it twice and still have an even number then the number is divisible by 8. E.g. 5208 Last 3 digits 208. Halve it 104, half again 52. Still even. So whole number is divisible by 8.
Divisible by 9 : Add the digits in the number. If the answer is divisible by 9 so the number. E.g. 5634 5+6+3+4 = 18 which divides by 9. So the number is divisible by 9
Divisible by 10 : Numbers end in 0 E.g. 13460

A Harder one:
E.g. Cancel down <u>168</u> to its lowest from.
                    252
          The point to this one is to firstly find ANY number that divides into both and take it from there. Since both numbers are even, let's start there:
<u>168</u> ÷ 2 = <u>81</u>   Now both numbers are divisible by 3 (can you see why?)   <u>81</u> ÷ 3= <u>27</u>
252 ÷ 2   126   126 ÷ 3   42
 and divide by 3 again.
So <u>27</u> ÷ 3 = <u>9</u> This is as far as you can go.
   42 ÷ 3   14

          The point here is that you need only find a number to divide exactly into both. You will eventually get to the right answer anyway.

A few more examples:
   Cancel down:

1.   <u>315</u> to its lowest from. Answer = <u>61</u>
     405   81

2. $\underline{24}$ to its lowest from.   Answer = $\underline{6}$
    52   13
3. $\underline{288}$ to its lowest from.   Answer = $\underline{32}$
    27   3

Teaching Points

1. The most important part of learning fractions initially, are the three areas mentioned at the start. To know how big a fraction is; to know which fraction is bigger and to be able to cancel down.
2. Most students get bogged down on arithmetic with fractions. To overcome student's natural fear (hatred?) of fractions, you should only focus on a few points and leave the rest. That is, focus on the size of fractions, comparing fractions, converting fractions and cancelling down.
3. However, to be able to multiply fractions is a useful skill for later Maths topics. Adding, Subtracting and dividing, not so much.

Extension Work

Testing which fraction is bigger

You could devise games for the students whereby you bring in say a chocolate bar (or something similar) and challenge them to point to where they think you should cut the bar to cut off, say $^1/_5$. If they are close, they get to keep the fraction. Similar games could be devised by getting them to select the bigger fraction of two fractions. Then using, say plasticine, cut them into the fractions to show which is bigger. The student with the most correct answers wins.

Converting Fractions into Mixed Numbers and decimals

Firstly, it should be emphasised that any answer left as a fraction is OK. $^{13}/_4$ is a perfectly acceptable answer. However in some practical cases, it may be necessary to convert the fraction into a mixed number or decimal. Again, the shortcut to this is to divide the bottom number into the top. For a mixed number, the remainder is the top of the fraction.

E.g. Convert $^{13}/_4$ into a mixed number. 13 ÷ 4 = 3 remainder 1. So the answer is 3 ¼

As a decimal, $^{13}/_4$ = 13 ÷ 4 = 3.25

Adding and subtracting Fractions

Again, I look at this process as an extension as you really only meet this again when dealing with algebraic fractions. Most students find this section hard and tedious. The practical applications of adding and subtracting fractions is really minimal and most calculators have fraction buttons which would provide the answer anyway. (A good extension exercise is showing students how to use fractions on a calculator). However, as a mathematical exercise, it is useful for your more able students to know how to add and subtract fractions.

The rule is simply that you can only add or subtract the top number of fractions and ONLY when the bottom numbers are the same.

E.g. $\frac{2}{9} + \frac{5}{9} = \frac{2+5}{9} = \frac{7}{9}$ Note: You only add (or subtract) the top numbers. The bottom number stays the same.

What happens if the bottom numbers are NOT the same.

E.g. $\frac{3}{4} + \frac{5}{12} = ?$

Step 1: Make the bottom numbers the same. How? Multiply top and bottom (equivalent fractions) of one or both fractions until both bottom numbers are the same.

E.g. $\frac{3 \times 3}{4 \times 3} + \frac{5}{12} = \frac{9}{12} + \frac{5}{12}$

Step 2: Add (or subtract) the top numbers only. The bottom number stays the same.

E.g. $\frac{9}{12} + \frac{5}{12} = \frac{9+5}{12} = \frac{14}{12}$ If we simplify this, we get $\frac{7}{6}$

A harder example:

E.g. $\frac{2}{7} - \frac{1}{6} = ?$ Step 1: We need to get the bottom numbers the same. An easy way of doing this is simply multiply the two bottom numbers. Then to keep the fractions the same, we need to 'cross multiply the top numbers. So:

E.g. $\frac{2}{7} - \frac{1}{6} = \frac{2 \times 6 - 1 \times 7}{7 \times 6} = \frac{12 - 7}{42} = \frac{5}{42}$

As a formula:

$$\frac{a}{b} + \frac{c}{d} = \frac{a \times d + b \times c}{b \times d}$$

Multiplying Fractions

Multiplying fractions might be a useful one to incorporate into the main part of fractions as it comes into play when calculating volume in measurement. Also when you are dealing with algebraic fractions. However, the process for multiplying is quite simple. You multiply the top numbers and then the bottom numbers.

E.g $\frac{2}{7} \times \frac{5}{6} = \frac{2 \times 5}{7 \times 6} = \frac{10}{42}$ Cancelling down we get $\frac{5}{21}$

A second example

E.g $\frac{3}{10} \times \frac{5}{9} = \frac{3 \times 5}{10 \times 9} = \frac{15}{90}$ Cancelling down we get $\frac{1}{6}$

Notice in the second step we combine the problem into one single fraction and considering that a fraction is really a divide, we could have cancelled down ANY number on the top with ANY number on the bottom **before we multiplied**. This:

$$\text{E.g } \frac{\overset{1}{3}}{\underset{2}{10}} \times \frac{\overset{1}{5}}{\underset{3}{9}} = \frac{1 \times 1}{2 \times 3} = \frac{1}{6}$$

So the method for multiplying fractions is:

Cancel down any number on top with any number on the bottom. Then multiply the top and bottom numbers.

$$\text{Example: } \frac{\overset{3}{6}}{7} \times \frac{9}{\underset{10}{20}} = \frac{27}{70}$$

NOTE: **Multiplying a fraction by a whole number.** The whole number is always on the top. This is simply because it could be written as that number over 1.

$$\text{Example: } \frac{6}{7} \times \overset{2}{14} = 12$$

Dividing fractions

Remember divide is the opposite of multiply. So the trick here is to **turn the second fraction upside down and multiply.**

$$\text{Example: } \frac{5}{8} \div \frac{9}{20} = \frac{5}{8} \times \frac{\overset{5}{20}}{\underset{2}{9}} = \frac{25}{18}$$

Decimals

Decimals can be confusing to some students. Knowing where to put the decimal point in a calculation can prove to be tricky. Most students will do decimal calculations on a calculator. Let's face it, even you and I would do the same. However, students can still get the wrong answer.

Rather than teach decimals as a separate topic, you might like to take some minutes in one of your lessons to just emphasise a few basic points about decimals.

Estimation

The most important aspect of doing arithmetic with decimals is estimation. No matter how you do the problem, it is handy to have some idea about the answer.

The trick is simply to pick 'nice' numbers that are close to the ones you are using. How 'nice' will depend on the individual. But they are trying to pick numbers that will allow them to calculate (estimate) the answer in their head.

Example 2.34 x 4.67 = ?

'Nice' numbers that are close are 2.34 -> 2 and 4.67 -> 5. Now 2 x 5 = 10. So the actual answer will be close to 10

I got this on my calculator 2.34 x 4.67 = 10.9278. (which is near 10)

Another example:

12.6 x 21.7 = ? Suppose we did this by pen and paper (the traditional way). While there are tricks to knowing where to put the decimal point in the answer, the best approach is to estimate

12.6
21.7 x (ignore the decimal point)
882
1260
25200
27342   Now we estimate the answer. 12.6 is near 10 and 21.7 is near 20. 10 x 20 = 200. So our final answer is 273.42

Notes:
1. We usually pick a multiple of 10 as our 'nice' number. However the further away the nice number is, the further away the actual answer is.
2. There are cases where moving the decimal point is a useful trick.
   For instance, find 0.63 ÷ 0.09 = ? Estimating in this case may not be so useful. So it is still a good idea to show the trick of moving the decimal point.
   In this case move the decimal point the same number of places for each:
   0.63 ÷ 0.09 = 63 ÷ 09 = 7

Fractions and decimals

Remember a fraction is a divide and a divide is a fraction.

E.g. $^1/_4 = 1 \div 4 = 0.25$

Each decimal place represents a power of 10. So 0.567, which has 3 decimal places, can be expressed as 567/1000 . Add a '0' for each decimal place.

E.g. 0. 43 = 43/100

Recurring decimals

Students need to be aware that recurring decimals exist and that they represent fractions (rational numbers)

E.g. What is $^1/_6$ as a decimal?

Remember a fraction is a divide and a divide is a fraction.

So $^1/_6 = 1 \div 6 = 0.1666666.....$

NOTE: The point to stress is that any fraction will convert to a decimal with either a **set number of decimal places OR recurring decimals**

E.g. 1. What is $^2/_5$ as a decimal? $= 2 \div 5 = 0.4$

2. What is $^6/_7$ as a decimal? $= 6 \div 7 = 0.857142857142....$

Teaching points

1. Try to encourage students to estimate the answer before calculation. Even using a calculator, students can still get it wrong but they have an in-built belief that the calculator is always right.
2. It is worth showing carious tricks about moving the decimal point in multiply and especially divide but it can get confusing for some students. That is why I push the idea of estimation.
3. Coupled with estimation is the concept of size. It is surprising the number of students who think that 0.098 is bigger than 0.123. Some reminding about size is also handy.
4. It may also pay to remind students about the idea of rounding off. I have assumed that it is common knowledge for secondary students but it is worth reinforcing.

Extension Activities

1. Non recurring decimals – Irrational numbers
   A famous non recurring decimal in Maths is π.
   π = 3.1411592... The decimal numbers never repeat nor is there any pattern in them.

How is it calculated? π is defined as the ratio of a circumference of a circle to its diameter. A way you could measure it is to simply get a piece of string and wrap it exactly around any circle you might find (jam jar, hoop etc.) Then measure the length of the string and divide it by the diameter of the circle. Do this for several different circles and average the results. This should give a good approximation to π. However it is only as good as your measuring.

A more mathematical way of calculating π is as follows:
$\pi = 4/1 - 4/3 + 4/5 - 4/7 + 4/9 - 4/11......$ The more terms you add in, the more accurate the answer. To get a large number of decimal places, mathematicians use super computers. In all their calculations, π has never repeated or ended the decimal places.

Other famous irrational numbers include 'e' (Euler's number), φ (the golden ratio) and $\sqrt{2}$. It is a useful exercise to investigate the origins of these numbers.

2. Scientific notation
    In Science, especially, they deal with very large and very small numbers. For instance, the number of atoms in 12 grams of carbon is approximately 602,300,000,000,000,000,000,000 atoms. This number is quite impractical to use in calculations, so scientific notation was invented.

    A number expressed in scientific notation is:
    Number between 1 and 10 X power of 10
    So the number above would be written as 6.023×10^{23}. This is more manageable.
    To write any number down in scientific notation:
    Step 1: Move the decimal point until it is between the first and second digit.
    Step 2: the number of places you moved is the power of 10 to multiply by
    Step 3: if the number is bigger than 1 originally, the power of 10 is positive. If it is less than 1 initially, the power of 10 is negative.

    Examples:
    1. 245,000 – (decimal point is at the end if not showing)

        2 4 5,0 0 0. $= 2.45 \times 10^5$
    2. 0.000289
        0.0 0 0 2 8 9 $= 2.89 \times 10^{-4}$

NOTES:

1. Most calculators will automatically convert very large or small numbers into scientific notation. Try this on your calculator $234,172^{456} = 3.22$ e+2448 (which means 3.22×10^{2448})
2. Numbers in scientific notation can be multiplied and divided like normal numbers (just remember to add or subtract the powers of 10). This is a good exercise to see if students really understand scientific notation
    Examples

1. $2.3 \times 10^4 \times 3.4 \times 10^5 = 7.82 \times 10^9$
2. $5.6 \times 10^3 \times 4.1 \times 10^7 = 22.96 \times 10^{10} = 2.296 \times 10^{11}$
3. $2.5 \times 10^5 \div 5.6 \times 10^3 = 0.446 \times 10^2 = 4.46 \times 10^{-1}$

Measurement

The topic of measurement should really focus on one main objective. That is, to be able to identify just how long or big something is from the measurement. For instance, the length of a window say, is 1.2m. Students should come to appreciate just how long that is. So when they are calculating measurements, they have an idea of just how big things are. The ability to calculate area say, is something that is used in specific fields and probably not so much in general everyday use.

Students tend to get bogged down very quickly with the myriad of formulae that exist in measurement. Having a sense of proportion gives them something to fall back on. The alternative method for measurement is to try and reduce the number of formulae to just a few. In this way students are not so overwhelmed.

Length

From above, the first objective that students should aim for is to have something 'concrete' that shows them just how long a measurement is.

Get the students to compile a short list of things they are familiar with that show them just how long each main unit of measurement is. For instance, this list:

Name	Short Name	Length
Millimetre	mm	Thickness of my thumbnail
Centimetre	cm	Width of my Finger
Metre	m	Distance from my chest to the floor
Kilometre	km	Distance from school to my house

NOTES:

1. The actual lengths can be really anything that they can identify with. The importance is that they when they get an answer of say, 20cm, they have some idea of just how long that is.
2. Best to stick to just the main metric units (above). Other units (such as micrometre, nanometre, Astronomical Unit) are for specialised fields and could be used as an extension. Also empirical units might be used as an extension as the building industry, for instance, still use them.

The main topic of length focusses on Perimeter. The formulae is simply the **sum of all the sides**

E.g. Find the perimeter of this paddock

60m

20m

Answer = 60 + 20 + 60 + 20 = 160m

There are several formulae for different straight sided shapes but all of them are variations on the sum of the sides.

NOTE: Perimeter is a length. So the units of measurement are units of length (metre, centimetre, kilometre etc.)

Area

Again, a knowledge of just how big each unit of area is, can help. Here is a similar table for area:

Name	Short Name	Area
Square Millimetre	mm^2	Area of the tip of a felt tip pen
Square Centimetre	cm^2	Area of my little fingernail
Square Metre	m^2	Area of a school desk
Square Kilometre	km^2	Area of Melbourne CBD.

Area, of course focusses on shapes. Each different straight sided shape has its own formula for its area. However, most are variations of the same shape. You can boil it down to just two basic ones:

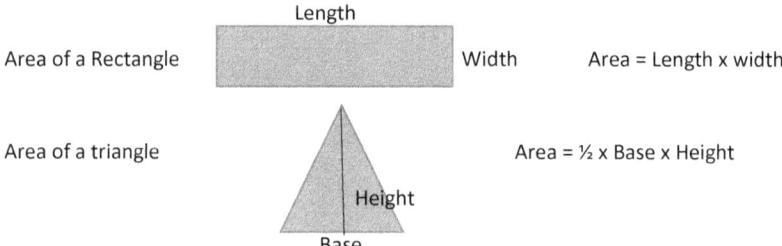

Area of a Rectangle   Length / Width   Area = Length x width

Area of a triangle   Area = ½ x Base x Height

Height / Base

NOTE: You can show that the area of a triangle is ½ x Base x Height by splitting a rectangle into two triangles.

That should be enough.

E.g. Find the area of:

1.

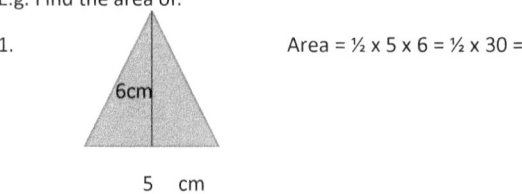

6cm

5   cm

Area = ½ x 5 x 6 = ½ x 30 = 15 cm^2

2. (Here is why you only need the 2)

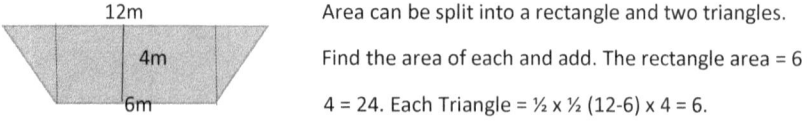

12m

4m

6m

Area can be split into a rectangle and two triangles.

Find the area of each and add. The rectangle area = 6 x 4 = 24. Each Triangle = ½ x ½ (12-6) x 4 = 6.

So the total area is 24 + 6 +6 = 36 m^2

NOTE: Even if the two triangles were not the same, the answer would be the same. If we call one base B1 and the other B2, then the area would be 24 + ½ x B1 x 4 + ½ x B2 x 4

= 24 + 2 x B1 + 2 x B2. But we do know that B1 + B2 must be 6 as the top length is 12. So 2x B1 + 2 x B2 must be 12. So the area is 24 + 12 = 36 as before.

This last example shows that shapes such as the trapezium, parallelogram, rhombus etc. can all be split into either rectangles and/ or triangles. So students really need only remember two formulae.

Surface Area

The only fact that students need know about surface area is that it is the area of all the surfaces of a solid.

E.g. Find the surface area of this pyramid:

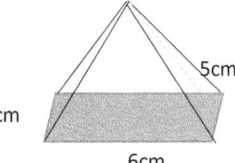

4cm

5cm

6cm

Surface area = Area of base = 6 x 4 = 24

+ 2 long faces = 2 x ½ x 6 x 5 = 30

+ 2 short faces = 2 x ½ x 4 x 5 = 20

= 24 + 30 + 20 = 74 cm^2

Volume

Again, having some idea of the size of volume can help. Here is a table of relative sizes (Note the extra units included):

Name	Short Name	Volume
Cubic Millimetre	mm^3	Volume of the tip of a felt tip pen
Cubic Centimetre	cm^3	Volume of a pea
Cubic Metre	m^3	Volume of a fridge
Litre	L	Standard milk container
Millilitre	ml	= 1 cm^3. Liquid in an eyedropper
MegaLitre	Ml	= 1000000l. The volume of an olympic sized swimming pool is about 3Ml

Note:

1. The km^3 is not a widely used measurement so can be ignored somewhat
2. The liquid measurements are used in cooking and general water usage and thus are quite common. E.g. Your water usage bill is measured in Megalitres.

There are several different types of 3D solids which each can have a formula for the Volume. However, you really only need two:

Prism – The identical shape at top and bottom, connected by equal straight sides.

Volume = Area of the base x Height

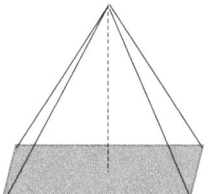

Pyramid- shape at the bottom, connected by equal straight sides to a point at the top.

Volume = 1/3 x Area of Base x Height

Example: Find the volume of:

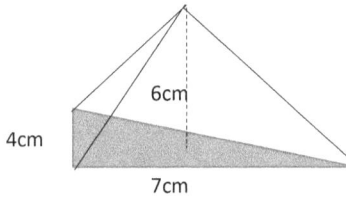

6cm

4cm

7cm

This is a pyramid. Volume = 1/3 Area of Base x Height

Area of Base = Area of a triangle = ½ x 4 x 7 = 14

Volume = 1/3 x14 x 6 = 28 cm³

Again, all other solid shapes (even cylinders and cones!) could be split into pyramids and/or prisms.

NOTE: A good exercise is to show with a tetrahedron, how 3 of them can be made into a prism, thus demonstrating the formula of a pyramid. Done by constructing 3 identical tetrahedrons and gluing their sides together to form a 'chain'. Then by simply bending the tetrahedrons together, you will get a prism.

Circular shapes

Treat these separately as they can be confusing, especially dealing with π.

Introducing π

Perimeter and area of a circle:

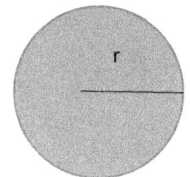

Perimeter = 2 π r   Area = π r²

NOTE: To help pick the difference remember that area involves square units, so its formula has the 'r squared' in it

What about π?

The ancient Mathematician, Archimedes, was one of the first to get an approximate number for π. He did it by the 'method of exhaustion'. He started with a circle of radius 1m. He then drew a pentagon, so it just fitted inside the circle and a second one just fitting outside the circle. From this he measured both perimeters. The perimeter of the circle had to be somewhere between the two. He repeated the process but each time used more and more sided polygons. The reason is imply that the more sides the polygon had, the closer it became to the circle (see below)

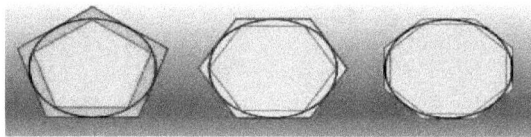

He got an answer of about 3.14

Facts about π:

- It is an irrational number. That is, the decimal places never end and never repeat. On a calculator, you might get an answer like 3.14159265358979323846264338332795 but the decimal places continue forever. Compare this with the number 1/3 which is 0.333333.. This number is rational because it repeats, even though it does not end.
- It has been calculated to over 10 trillion (10^{13}) digits by computers so far. Still no end or repetition. Also there are many advanced formulae that can calculate π.
- For practical purposes, use either 22/7 or 3.14 for any calculation

E.g. Find the area of a circle measuring 7m in diameter

Area = π r² = 22 x 7 x 7 = 22 x 7 = 154 m² (you will find most exercises use radii of multiple of 7.
           7   So using 22/7 for π should be your first choice)

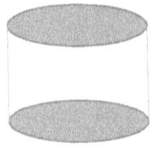

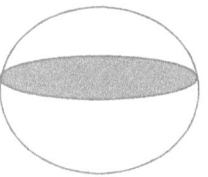

| Cylinder (Prism) | Cone (Pyramid) | Sphere |

The Volume of a pyramid is the volume of a prism (area of base x height) and the volume of a cone is the volume of a pyramid (1/3 area of the base x height). The Sphere is more tricky. It's volume is: **Volume = 4/3 π r³**

As for **surface area:**

Cylinder: Has two circles (top and bottom) and the sides folds out to a rectangle with its area being the height x perimeter of the circle **(2 x π r² + 2 π r x h)**

Cone: Has the circle base and the side folds out to a circle with a wedge cut out. The size of the wedge relates to the height of the cone. The radius of the circle wedge is the slant height and the arc length is the perimeter of the base circle. You may see from this that calculating the surface area of cones is something to avoid.

Sphere: again it simply has a formula: = **4 π r²**

Example: Find

1. Surface Area of a cylinder with base radius of 14cm and height of 6cm.

Surface area = 2 x π x 14 x 14 + 2 x π x 14 X 6 = 2 x $\underset{7}{\overset{2}{\underline{22}}}$ x 14 x 14 + 2 x $\underset{7}{\overset{2}{\underline{22}}}$ x 14 x 6 = 1232 + 528 = 1760 cm²

2. Find the volume of a cone with a base radius of 3.5 m and a vertical height of 9m

Volume = 1/3 area of base x height = $\underset{3}{\underset{2}{\overset{11}{\underline{1}}}}$ x $\underset{7}{\overset{1}{\underline{22}}}$ x 3.5 x 3.5 x 9 = 115.5 m³

NOTE: The sphere and the surface area of the cone are quite advanced for your average middle school student. You should only attempt them if you feel students have the confidence. Otherwise skip them. The important point here is not to overwhelm them with formulae.

Units of Measurement

I've shown examples of the units above. Best to stick to the metric system we commonly use. However, students to get bogged down in converting units of measurement. To get over this, there are some simple facts to keep in mind:

The prefixes have meaning: centi = 1/100, milli = 1/1000, mega = 1,000,000

For conversion, simply keep in mind for area you square the answer and for volume you cube it.

E.g. How cm^2 in 3m^2 ? From our prefixes there must be 100 cm = 1m. For area we square this (i.e. 100 x 100 = 10000). So 3m^2 = 30000cm^2.

Teaching Points

1. Focus on the actual size of each measurement. This has more practical value for students than say finding the area of a triangle.
2. Keep the number of formulae to a minimum to reduce confusion.
3. Stick with metric units. However, looking at imperial units might be a good extension exercise. (American and UK measurements are in imperial).

Extension Work

Find the area of shapes

A good activity to try is firstly to collect a set of small wooden straight sided shapes. They can be as complex or as simple as you like. Your woodwork department might be able to help. Try to collect about 10 of them. Next carefully measure the perimeter and area of each and keep these aside. The activity then involves splitting your class into pairs and give each pair a ruler and one of the shapes. Their challenge is to measure the perimeter and area within a short time period (say 2 minutes). Once the time is up, they record their results and pass the shape onto the next pair (It is a good idea to put a number on each shape to help identify them). Once each pair has measured all the shapes, get them to hand in their results. You can compare the results and perhaps, award a prize for the pair who have the most correct.

Construct 3d models

Spending a lesson or two, constructing 3D shapes , not only gives the students a sense of their size and shape but also is a welcome break from doing repeated maths problems. Try to go for the more unusual 3D shapes, not just the 5 Platonic solids. There are many websites and books that have detailed instructions on how to make them. All you need is thin cardboard, scissors and glue.

Proof that the volume of a pyramid is 1/3 the volume of a prism.

Construct the following shapes out of thin cardboard:

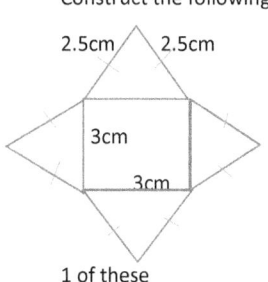

1 of these

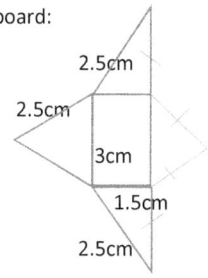

4 of these (sides with markers are all the same length)

Fold them up to make a large pyramid and 4 smaller ones. The 4 smaller ones can be put together to form two larger pyramids the same size as the first one. The shapes can them be shifted around to form a cube thus showing that 3 pyramids make a cube!

A pyramid ring

For this one you will need some thin cardboard, ruler, protractor, scissors and glue. In the diagram below, each triangle is an equilateral triangle. If you draw the diagram shown on say thin card and cut out the shape, then lightly score along each line (the black lines, score on the front side and the red lines, score on the back). After folding the lines, join (and glue) the tabs to the sides with the same number. I.e. Join Tab 1 to side 1 etc.. You will have constructed a set of pyramids in a ring that spins around forever!

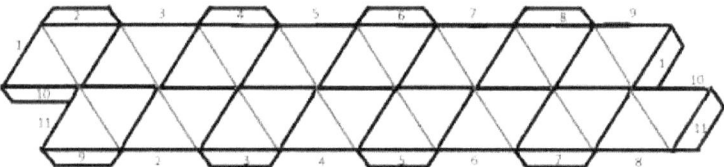

A practical experiment to determine π.

On a large sheet of paper rule parallel lines 2cm apart. Cut 10 matches so that they are each 1cm long. Holding the matches in your hand about 1m above the centre of the paper, drop the matches on the paper and record how many matches have crossed or touched a line. Repeat this several times and average your results. The number of matches dropped divided by the number of matches crossing the lines should be very close to π.

Percentages

Percentages can be tackled by using 3 simple rules:

1. **A Percentage is a fraction out of 100**
2. **A fraction is a divide and a divide is a fraction**
3. **Multiplying by a whole number is the same as multiplying by the fraction over 1**

I'll demonstrate with some examples:

1. Find 20% of $45

"A Percentage is a fraction out of 100". Whenever you see '%' replace it with '/100'. The only other point here is that 'of' means 'x' (remember 'bodmas'. The 'o' means 'of')

So: 20% of 45 = 20/100 x 45   and **"Multiplying by a whole number is the same as multiplying by the fraction over 1".**

So 20/100 x 45 = $\dfrac{20}{100} \times \dfrac{45}{1} = \dfrac{20}{100} \times \dfrac{45}{1} = \dfrac{45}{5} = \$9$

2. 5 students in a class of 35 are male. What % of the class are male?

"A fraction is a divide and a divide is a fraction". So 5 out of 35 is really 5/35. Express this as a %. **"A Percentage is a fraction out of 100".**
Hence 5/35 = ?/100 (This is a simple linear equation – get ? by itself)

$\dfrac{5}{35} = \dfrac{?}{100}$

$\dfrac{5 \times 100}{35} = ?$ $\dfrac{5 \times 100}{35} = ?$ (reduces to $\dfrac{5 \times 100}{7}$)

? = 100/7 (which could be good enough for a final answer) OR **"A fraction is a divide and a divide is a fraction".** So ? = 100 ÷ 7 = 14.285%

3. An investment of $2000 increases by 15% in the first year. How much is the investment worth now?
There are several ways you work this out but using our rules:
First find 15% of 2000
= $\dfrac{15}{100} \times \dfrac{2000}{1}$ (by rules 1 & 3)

= $\dfrac{15}{100} \times \dfrac{2000^{20}}{1}$

= $\dfrac{15}{100_1} \times \dfrac{2000}{1}$

= 15 x 20 = 300
So the investment has increased by $300 giving a total of $2300

That's really about it for basic percentages

Teaching Points

1. From above you can see that percentages can be handled by reinforcing those 3 basic rules.
2. There are usually at least two different ways of solving % problems. Try to stick to just one way for most students to avoid confusion.

Applications of Percentages (Extensions)

Once the basic calculations are mastered, there only remains to investigate common applications of percentages. The majority of these are concerned with finance. Here are some:

1. **Simple Interest**

   This relates to investing some money at a fixed rate (%) per annum (per year). Example 3 above is an example of such an investment. The point here is that the amount is fixed every year.

   Example:

   A woman invests $5,000 at 3%p.a. simple interest over 4 years. How much has she made in that time?

   Find 3% of $5,000 = 3/100 x 5000/1 =3/1 x 50/1 = $150

   Since the investment is for 4 years, she makes $150 x 4 = $600

2. **Compound Interest**

   Using the example above, the idea of compound interest is that after the first year, the woman actually has $5,000 + $150 = $5,150. So the 3% interest for the second year should really be calculated on $5,150 not the original $5000. This is compound interest.

   Example:

   What is the final amount of an investment of $20,000 invested for 4 years at 5% compound interest?

   You could do this manually by calculating the total amount for each year, thus:

   Year 1: $20,000 + 5% of $20,000 = $20,000 + 5/100 x 20,000 = $20,000 + $1,000 = $21,000

   Year 2: $21,000 + 5% 0f $21,000 …..

   There is a formula that calculates it much more quickly:

   $A = P(1 + r)^n$ where A = Amount accumulated; P = the original principle; r = the % rate and n = number of years of investment. Using this we have:

   $$A = P(1 + r)^n$$
   $$= 20000 \times (1 + 5/100)^4$$
   $$= 20000 \times (1 + .05)^4 \quad \text{(remember a fraction is a divide...)}$$
   $$= 20000 \times 1.05^4$$
   $$= 20000 \times 1.21550625$$
   $$= \$24310.13$$

NOTE: a good mathematical exercise is to investigate where that formula comes from.

3. **Discount, Commission, Profit and Loss**
    All of these are really just practical problems involving %, much the same as our original examples. For instance:
A camera selling for $900 is discounted by 10%. What is the new price?
Find 10% of $900 = $90. So the new price is $900 - $90 = $810
(You might care to approach it in a different way by assuming that the original price of $900 represents 100%. A discount of 10% means that the new price is then 90% of the original. 90% of $900 = $810)
A special example:
A seller gets 15% commission for all her sales. Last year she earned $40,000. How much did she sell?
The problem really means that 15% of sales = $40,000. What therefore is the amount of sales? i.e. what is 100% if 15% = 40000?
Solution
15% = 40000 ->   15/100 x ? = 40000/1 (by rules 1 and 3)
(get ? by itself)

$$\frac{15}{100} \times ? = \frac{40000}{1} \times \frac{100}{15}$$

? = $\frac{4000000}{15}$ = $266,666.67

Integers (Positive and Negative numbers)

In dealing with positive and negative numbers, students tend to get confused about whether you actually add or subtract the numbers and whether the signs are positive or negative. We can simply this somewhat.

Why have integers?

Positive and especially negative numbers, exist in everyday usage. Here are 3 examples:

1. The temperature at Mt. Hotham last week was -4°C.
2. The depth of the pacific ocean is 10,911m below sea level or -10,911m.
3. The profit for the company last year was -$2,300,000 (i.e. they made a loss)

The traditional approach to learning about integers is to use the number line. This is good for an introduction but it really should only serve to help explain what integers (in particular arithmetic with integers) actually do.

Number Line:

The main point to emphasise at this stage is that numbers increase to the right. So -5 is less than +3. Hence for adding numbers, you move to the right and subtract move to the left.

Also that +3 say, is really just 3.

Adding and Subtracting Integers

Using the number line we will look at the following examples:

Find:

1. $^{+}2 + {}^{+}3 = ?$ On the number line this simply means start at $^{+}2$ and move to the right (That's the + in the middle) $^{+}3$ places. Answer $^{+}5$.

Note: Adding means move to the right. So subtracting means move to the left.

2. $^{+}3 - {}^{+}6 = ?$ On the number line this simply means start at $^{+}3$ and move to the left (That's the - in the middle) $^{+}6$ places. Answer $^{-}3$.

Note: moving $^{+}6$ places must be the opposite to moving $^{-}6$ places

3. $^{-}1 + {}^{-}4 = ?$ On the number line this means start at $^{-}1$ and move to the right (That's the + in the middle) $^{-}4$ places. Which really means move back 4 places. Answer $^{-}5$.
4. $^{-}2 - {}^{-}5 = ?$ On the number line this means start at $^{-}2$ and move to the left (That's the - in the middle) $^{-}5$ places. Which really means move forward 5 places. Answer $^{+}3$.

OK. Is there an easier way of doing this? Yes there is!.

Let's firstly have a look at the signs in the middle.

1. $^+2 + {}^+3 = {}^+5$ - the answer actually came from moving to the right on the number line. I.e. adding. So $+ {}^+ = +$.
2. $^+3 - {}^+6 = {}^-3$ – Answer found by moving left. I.e. subtracting. So $- {}^+ = -$
3. $^-1 + {}^-4 = {}^-5$ - Answer found by moving left. I.e. subtracting. So $+ {}^- = -$
4. $^-2 - {}^-5 = {}^+3$ – Answer found by moving right. I.e. adding. So $- {}^- = +$

This leads to a simple rule:

For double signs, replace like signs with + and unlike signs with –

Example:

$^-1 - {}^+5 = -1 - 5 = -6$

Now let's look at the number in the answer. Again going back to our original examples:

1. $^+2 + {}^+3 = +2 + 3 = +5$ - Ignoring the signs for the moment, the answer was found by adding the numbers.
2. $^+3 - {}^+6 = +3 - 6 = -3$ – Answer found by subtracting the numbers.
3. $^-1 + {}^-4 = -1 - 4 = -5$ - Answer found by adding the numbers.
4. $^-2 - {}^-5 = -2 + 5 = {}^+3$ – Answer found by subtracting the numbers.

This leads us to our second rule:

If the signs in front of the numbers are the same, then add. If they are different then subtract.

Example:

$-3 + 7 = ?$ Signs are different so the answer is going to be $7-3 = 4$. But will it be +4 or -4?

Again checking our original examples, have a look at the sign of the 'larger' number and the sign of the answer:

1. $+2 + 3 = +5$ - 'Larger' number is 3 and +. Answer is +
2. $+3 - 6 = -3$ - 'Larger' number is 6 and -. Answer is -
3. $-1 - 4 = -5$ - 'Larger' number is 4 and -. Answer is -
4. $-2 + 5 = {}^+3$ - 'Larger' number is 5 and +. Answer is +

This leads us to our third rule:

The sign of the larger number is the sign of the answer

Examples:

1. $^+2 + {}^-6 = ?$ Double sign in the middle and different signs. So replace with –
   $^+2 + {}^-6 = +2 - 6$ Signs in front of the numbers are different so subtract (6-2 = 4). Sign of the larger number (6) is -. So answer is -.
   $+2 - 6 = -4$
2. $8 - 10 =$ The 8 is really +8. Hence the signs are different. So subtract (10-8 = 2). The sign of the larger number (10) is -. Answer is –
   $8 - 10 = -2$
3. $-7 - 9 =$ Signs are the same so add. Sign of larger number (9) is -. So answer is:
   $-7 - 9 = -16$

So, in summary to add and subtract integers, you have 3 steps:

1. **Any double signs convert to single (same sign converts to +, different to -)**
2. **If the signs in front of the numbers are the same, add. If not, subtract.**
3. **Sign of the larger number is the sign of the answer.**

Multiply and Divide Integers

Example:

1. $^-4$ x $^+2$ = ? As $^+2$ is really just 2, this could be written as $^-4 + ^-4$ = -4 − 4 = -8 (from above)

So $^-4$ x $^+2$ = $^-8$

2. $^+4$ x $^+2$ = 4 x 2 = 8 = $^+8$
3. $^+4$ x $^-2$ = $^-2$ x $^+4$ = $^-2$ x 4 = $^-2 + ^-2 + ^-2 + ^-2$ = $^-2 − 2 − 2 − 2$ = $^-8$
4. $^-4$ x $^-2$ =-(4 x $^-2$) =-(-8) = $^+8$

These examples show us that:

For multiply (and divide as well), you multiply (or divide) the numbers. If the signs are the same, then answer is +, if not the answer is -.

Examples

1. $^-6$ x $^-4$ = ?   6 x 4 = 24, signs are the same so answer is +

    $^-6$ x $^-4$ = $^+24$

2. $36 ÷ ^-9$ = ?   36 ÷ 9 = 4, signs are different (36 is really $^+36$), so answer is −

    $36 ÷ ^-9$ = $^-4$

Teaching Points

1. For all the rules, same signs = +, different signs = -
2. If there is no sign in front of the number it is positive E.g. 3 = +3
3. As you can see, only use the number line as an introduction. The rules above are much easier to deal with and relate much better to ordinary arithmetic.
4. Another concept that may not be apparent is that 'positive' and 'add' can really mean the same thing (and 'negative' and 'subtract' also mean the same thing). You can change between them whenever it suits you. So $^+3$ is the same as +3, say. Hence it does not matter whether you write $^+3$ or +3. This becomes useful when dealing with algebraic terms later on.

Extension Activities

Integer Adder

Copy the following diagram onto a piece of cardboard.

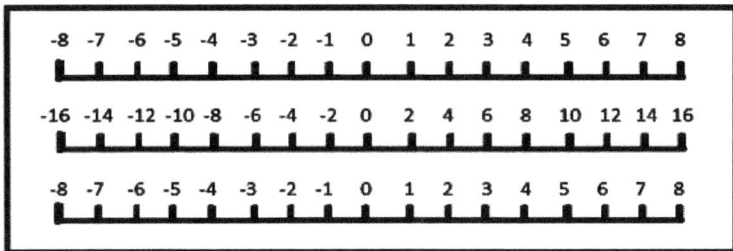

How to use:

To add any two integers, hold your ruler (or any straight side) across the card so that the first number lines up with the top scale and the second number lines up with the bottom scale. The answer is on the middle scale.

Examples

1. -6 + -2   Answer from below is -8

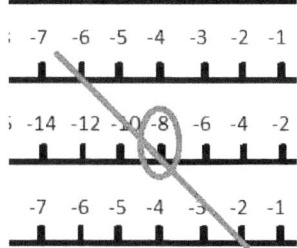

2. -3 + 7   Answer from below is 4

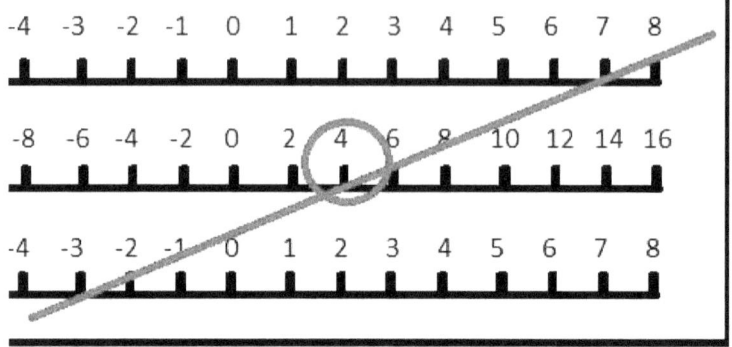

NOTE: For subtraction, Use the middle scale number minus the bottom scale number. The answer is on the top scale.

Example: -4 - -3 Answer is -1 thus:

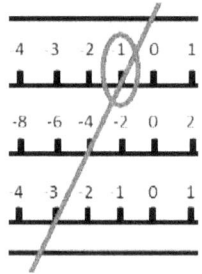

Graphs and Co-ordinates

Once students have mastered the basic idea of integers, this is an opportune time to move on to the topics of **Co-ordinate Geometry** and **Linear Graphs.**

Indices

When you multiply a number by itself, you can express it in 'index form'.

E.g. $3 \times 3 = 3^2$ (the 2 means it is multiplied by itself 2 times). This is called index notation.

This is called the BASE NUMBER 3^2 This is called the INDEX or POWER (The plural of Index is Indices!)

Why bother doing this? The answer is because of Algebra. If we replace the number '3' with a pronumeral (say 'a'), which we do not know what number it represents, the index notation provides a useful short way of writing the expression.

E.g. $a \times a = a^2$

Other examples:

$2 \times 2 \times 2 \times 2 \times 2 = 2^5$

$a \times a \times a = a^3$

Multiplying terms in Index Notation

E.g. $a^2 \times a^3 = ?$

If we rewrite the terms in their expanded form, you can see what happens:

$a^2 \times a^3 = a \times a \ \times \ a \times a \times a = a^5$

The shortcut is simple. Just **add the powers**:

$a^2 \times a^3 = a^{2+3} = a^5$

Here are some other examples:

$B^4 \times B^5 = B^{4+5} = B^9$

$C^4 \times C^3 = C^{4+3} = C^7$

Note that the **Base Term DOES NOT CHANGE**. Also the **Base Terms MUST be the same** for this to work.

E.g. Let's jump back to numbers: $2^3 \times 3^4 = ?$

If we expand like we did above, we get:

$2^3 \times 3^4 = 2 \times 2 \times 2 \ \times \ 3 \times 3 \times 3 \times 3 = 2^3 \times 3^4$

Base Numbers have to be the same for the rule to work.

So, the rule is:

> ## To Multiply Index terms with SAME BASE you add the powers

E.g. (start with numbers) $3^5 \div 3^2 = ?$

From our work in Arithmetic, we should recall that 'all divides are fractions'. So we could rewrite this as a fraction:

$\dfrac{3^5}{3^2}$ = ?. Expanding the indices, we get

$\dfrac{3^5}{3^2}$ $\dfrac{= 3 \times 3 \times 3 \times 3 \times 3}{3 \times 3}$ Again, from Arithmetic, the bottom 3's cancel out with 2 of the top '3's

thus:

$\dfrac{\cancel{3} \times \cancel{3} \times 3 \times 3 \times 3}{\cancel{3} \times \cancel{3}} = \dfrac{3 \times 3 \times 3}{1} = 3^3$

The shortcut is therefore to **subtract the powers.**

$3^5 \div 3^2 = 3^{5-2} = 3^3$

Some Examples using Algebra:

$a^6 \div a^2 = a^{6-2} = a^4$

$b^7 \div b^5 = b^{7-5} = b^2$

In Algebra all divides are expressed as fractions. So the rule above is not really applicable.

E.g. $\dfrac{a^5}{a^2}$ = ?. Expanding the indices, we get

$\dfrac{a^5}{a^2} = \dfrac{a \times a \times a \times a \times a}{a \times a}$ Again, from above, the bottom a's cancel out with 2 of the top 'a's even though we do not know what 'a' is. So:

$\dfrac{\cancel{a} \times \cancel{a} \times a \times a \times a}{\cancel{a} \times \cancel{a}} = \dfrac{a \times a \times a}{1} = a^3$

Yes this looks exactly the same as the example from before. However, when we look at what has happened a little more closely, the 'a^2' on the bottom has **cancelled out** with 2 of the indices on the top.

i.e. $\qquad \dfrac{a^{\cancel{5}\,3}}{\cancel{a^2}} = a^3$

So the rule for dividing is:

Cancel out indices with same base by removing the term with the lower power and subtract that power from the other term.

Some examples:

1. $\dfrac{a^6}{a^4} = \dfrac{a^{6}2}{a^{4}} = a^2$

2. $\dfrac{a^2}{a^4} = \dfrac{a^{2}}{a^{4}\,2} = \dfrac{1}{a^2}$

The last example shows the advantage of this rule in that it doesn't matter whether you cancel on the top or bottom of the fraction. Also note that when the lower power term is cancelled, it is actually replaced by 1 (as you do in Arithmetic)

Some special cases

Negative Powers

Returning to numbers for a moment:
E.g. $3^2 \div 3^5 = ?$

If we apply our rule from before about subtracting the powers, then the answer is:

$3^2 \div 3^5 = 3^{-3}$

This example shows that negative powers do exist. But what are they really? If we apply our process from before and convert the problem to a fraction then we have:

$$\dfrac{3^2}{3^5} = \dfrac{3^{2}}{3^{5}\,3} = \dfrac{1}{3^3}$$

So 3^{-3} means $\dfrac{1}{3^3}$

As a rule: **Negative powers mean 1/ the 'positive power'**
Note: Unless students feel comfortable with negative numbers, they should rewrite any negative powers as 1/...

E.g.

1. $a^6 \times a^{-4} = \dfrac{a^6}{a^4} = \dfrac{a^{6}\,2}{a^{4}} = \dfrac{a^2}{1} = a^2$

2. $a^7 \times a^{-10} = \dfrac{a^7}{a^{10}} = \dfrac{a^7}{a^{10}\ 3} = \dfrac{1}{a^3}$ (This could be written then as a^{-3})

Zero Power

Zero powers are easy. Consider the following example:

E.g. $\dfrac{a^4}{a^4} = \dfrac{a^4}{a^4} = 1.$ (Anything divided by itself is always 1)

But we could again have solved this by using the rule of subtracting powers.

$$\dfrac{a^4}{a^4} = a^4 \div a^4 = a^{4-4} = a^0$$

So as a rule: $a^0 = 1$ (anything to the power 0 is always 1)

E.g. $\dfrac{b^2}{b^2} = \dfrac{b^2}{b^2} = 1.$

Teaching Points:

1. Use this topic as lead in to Algebra as the methods of simplifying expressions are really centred on simplifying indices.
2. Try to keep students away from negative powers if they are struggling. Use the '1/' rule to convert any negative powers.

Extension Activities

Fractional Powers

So far we have seen powers (indices), written as positive and negative whole numbers and 0. Do fractional powers exist and what would they mean? To answer this, we will firstly have a look at Square Roots

Square Roots

Square Roots deal with the power of 2 exclusively. We know that $3^2 = 3 \times 3 = 9$. We usually say '3 squared is 9'.

Square Roots ask the question "What number squared gives an answer of 9?".

Or $?^2 = 9$. In Maths we write this problem using a different notation, the square root sign. I.e. $\sqrt{9} = ?$ (Remember this means what number multiplied by itself gives an answer of 9.)

Now we know the answer is 3. $\sqrt{9}$ = 3. So here is a harder one:

$\sqrt{576}$ = ? To work this out manually, you could guess the answer. Say 10 x 10 = 100, no. 12 x 12 = 144, no etc. I hope you can see that it may take some time to guess the answer.

The easiest way to work it out is to use the $\sqrt{}$ button on your calculator.

$\sqrt{576}$ = 24 (Don't forget this means 24 x 24 = 576)

E.g. Here a few others

1. $\sqrt{2704}$ =52

2. $\sqrt{1444}$ =38

NOTE: It is important for students to keep in mind what square roots actually mean. Also this could be used as a lead in to Surds and Complex numbers

Using the example from before, $\sqrt{9}$ = 3, we could express this as an index, so:

$9^?$ = 3. To solve this can be done thus:

Rewriting this problem, we have simply:

9 = 3 x 3

From our basic rule for indices, this problem can also be rewritten as:

$9^1 = 9^? \times 9^?$ (Don't forget that 9^1 = 9!)

Using our basic index rule for multiplying, we get:

? + ? = 1

? = ½ (since both ? are the same)

So:

$\sqrt{9} = 9^{\frac{1}{2}}$ = 3

As a rule:

$$\sqrt{a} = a^{\frac{1}{2}}$$

From this we learn three things:

1. Fractional powers do exist.
2. The powers follow the same rules as before
3. Since all fractions can be written as decimals, then decimal powers exist as well. Try $3^{2.4}$ = ? on your calculator (Use the Power button) . $3^{2.4}$ = 13.9666..

NOTE: The power ½ was worked by starting with the power of 2. Other fractional powers could be worked out by using a similar method to above with different powers

Examples:

1. $a^{2.4} \times a^{3.5} = a^{2.4 + 3.5} = a^{5.9}$

2. $a^{3.2} \div a^{1.4} = \dfrac{a^{3.2}}{a^{1.4}} = \dfrac{a^{3.2}\ 1.8}{a^{1.4}} = \dfrac{a^{1.8}}{1} = a^{1.8}$

Pythagoras' Theorem

There are many proofs of Pythagoras' Theorem available on the internet and in books.

Here are a few:

1. Look up Pythagorean Theorem Water Demo on the internet – Shows the squares of the two shorter sides filled with water. When you tip it over, the water exactly fills the square on the third side
2. Using actual squares:

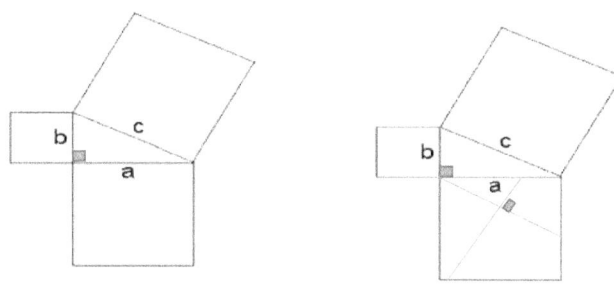

   Start by creating a square for each side. Then draw a line, parallel to side 'c' from the right angle to the edge of side 'a'. Then draw another line at right angles this one, also from one side of 'a' to the other. Cut out square 'a' along these lines. Also cut out square 'b'. The 4 shapes plus square 'b' should exactly fit over square 'c', thus:

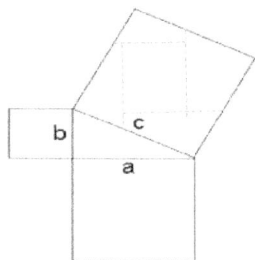

3. Creating a rectangle.
   Start with a right angled triangle and construct 3 similar triangles by multiplying each side by 'a', then 'b', then 'c', thus:

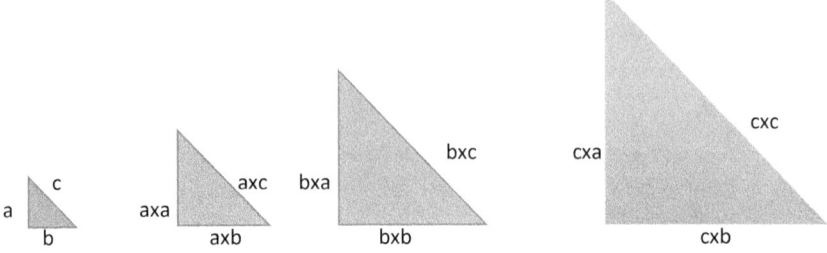

Place the 3 similar triangles together, thus:

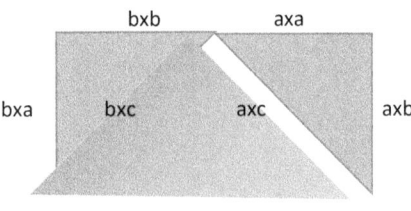

From the dimensions of the rectangle, you can see that cxc = axa + bxb or

$$a^2 + b^2 = c^2$$

Teaching Points:

1. There are several hundred different and elegant proofs of Pythagoras' Theorem. Most can be found on the internet. You may care to show a proof as a class exercise, depending on the students' ability but for simplicity sake, it is better to demonstrate with your own cut outs.

2. Some students have difficulty with accepting that the third side, the hypotenuse, is indeed the longest side. This can be easily shown by simply drawing a horizontal line. Then placing a set square (or any right angled side) at one end and get them to carefully rotate the square upwards. They will see that the side of the right angle will never form a triangle. The only way to do it is by making the sides shorter.

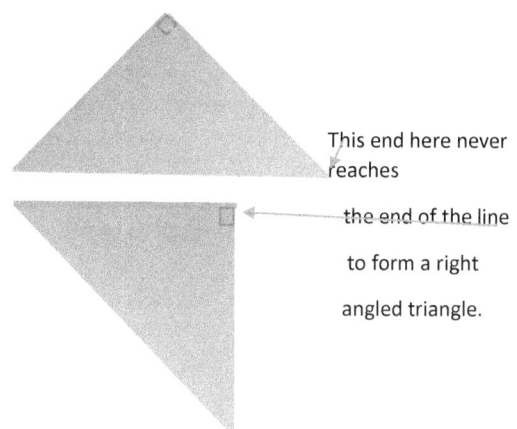

This end here never reaches

the end of the line

to form a right

angled triangle.

3. Proving Pythagoras' Theorem is probably secondary to students remembering how to use it

Using Pythagoras' Theorem

Examples

In each case find x:

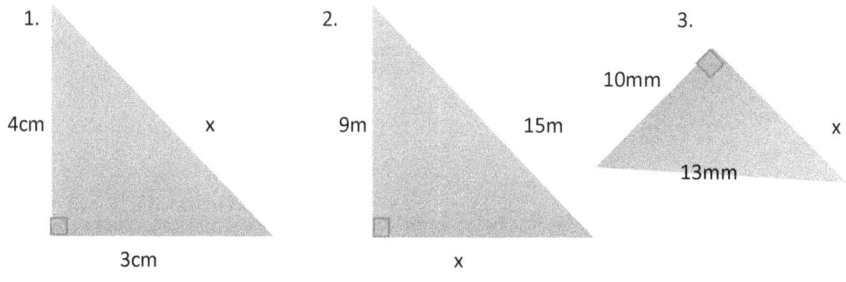

1. $a^2 + b^2 = c^2$
   $3^2 + 4^2 = x^2$
   $9 + 16 = x^2$
   $25 = x^2$
   $5cm = x$

2. $a^2 + b^2 = c^2$
   $9^2 + x^2 = 15^2$
   $81 + x^2 = 225$
   $x^2 = 225 - 81 = 144$
   $x = 12m$

3. $a^2 + b^2 = c^2$
   $x^2 + 10^2 = 13^2$
   $x^2 + 100 = 169$
   $x^2 = 169 - 100$
   $x = \sqrt{69}$ mm

NOTES:
1. Step 1 in any of these problems is to identify the hypotenuse-> Opposite the right angle.

2. Strictly speaking the $\sqrt{25}$ say, is really ±5 but since we are dealing with lengths, then the -5 can be ignored.
3. If the answer does not work out to be a nice whole number (like question 3), it should really be left in square root form (or perhaps simplified if you have done surds). This removes one extra complication from the problem.

Pythagorean Triples

It shouldn't take students long to realise that there are some answers that are nice whole numbers. These are called **Pythagorean triples**. Here are 3 examples:

3,4,5 $\qquad$ $3^2 + 4^2 = 5^2$

5,12,13 $\qquad$ $5^2 + 12^2 = 13^2$

7,24,25 $\qquad$ $7^2 + 24^2 = 25^2$

It is a useful exercise for students to memorise at least these triples above.

Also, any multiple of these triples is also a Pythagorean triple. For instance,

3,4,5 multiply each by say 2, you get 6,8,10. $6^2 + 8^2 = 10^2$

5,12,13 multiply each by say 3, you get 15,36,39. $15^2 + 36^2 = 39^2$

7,24,25 multiply each by say 4, you get 28,96,100. $28^2 + 96^2 = 100^2$

And so on..

Teaching Points

The main points for this section is for students to memorise Pythagoras' theorem and know how to use it in problems. A useful thing to remember are the Pythagorean triples and the fact that multiples of these are also Pythagorean triples. This should save them a lot of hassle when doing problems.

Extension Activities

Proof of Pythagoras' Theorem

As I said before there are 100s of proofs of Pythagoras' Theorem. Some involve, geometry, trigonometry and algebra. If students are familiar with these, it is an interesting exercise to get students to research and find out other proofs.

Pythagorean Triples

See if students can discover other Pythagorean triples (called primitive triples – 3, 4, 5 is a primitive triple, 6,8,10 is not). There are 16 primitive triples where all numbers are less than or equal to 100. Can you find them?

Pythagoras' Theorem in 3D

The theorem can be used on 3D diagrams. Here is an example:

A pencil is placed inside a cylindrical case so that it rests level with the top. If the case is 12cm high and the diameter of the base is 5cm, find the length of the pencil.

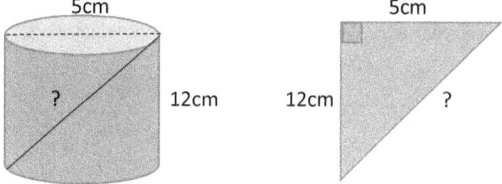

$a^2 + b^2 = c^2$

$5^2 + 12^2 = ?^2$ (Pythagorean triple)

$? = 13$ cm

Angles

This section is a very good example of the power of Mathematics to start from a simple fact and expand it into several complex rules.

Starting with a simple fact about angles: The straight line has an angle of 180°

180°

If we now cross that line with another, we create 4 new angles, thus:

a   b

c   d

By above, we can see that the angles a & b must add up to 180°.
So a + b = 180.

Also since the crossing line is also straight, angles a+c = 180° also.
Hence

a + b = 180
a + c = 180

So b = c.

Using the same argument you should be able to show that angles a and d are also the same.

In Maths, we use names to describe this. We call angles a and d **OPPOSITE** angles. Also angles which are the same are called **CONGRUENT** angles and angles which add up to 180° are called SUPPLEMENTARY.

So angles a and d are opposite and congruent and angles a and b are supplementary. But it is easier to show this via a diagram. We use little curves to represent each angle, thus:

The ones with the identical curve have the same value. We can also say that

 = 180°

Suppose we add a third line, but this one is parallel to the second, So:

(Notice that the parallel lines are indicated by the arrows). In simple terms, we have literally 'shifted' the intersection down the horizontal line a little. Hence we have the identical intersection as before. So the same rules are true:

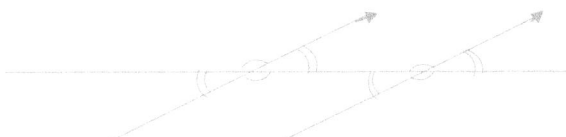

Example:
Find angles a and b

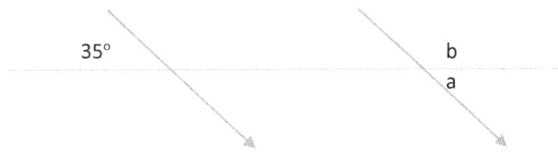

By the diagram above, a = 35° and b= 145° since a + b = 180°

Triangles

From our parallel lines diagram, if I add another line that crosses the intersection of two of the lines, I actually form a triangle, thus:

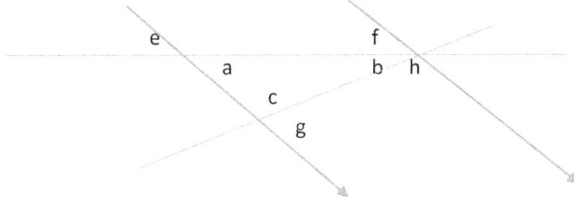

The angles inside the triangle, I have labelled a, b, c.
The angles f + b + h = 180° (angle of a straight line),
By the rules above, angles f = a, and h = c (Check with the parallel lines)
So angles   a + b + c = 180°
This gives us the rule that the angles in ANY triangle add up to 180°
(A further test is to draw ANY triangle, then draw a parallel line to any ONE of the sides so that it touches the opposite corner. You will then have redrawn the diagram above)

Example:
Find a

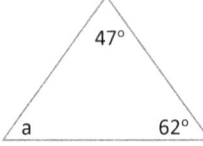

a + 47 + 62 = 180
a = 180 – 47-62
a = 71°

Types of Triangles

Triangles come in different varieties:

If no side in the triangle has the same length, it is called a **SCALENE TRIANGLE**:

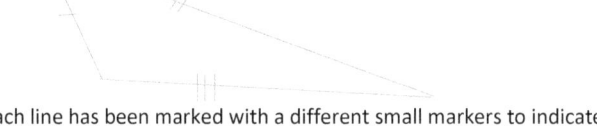

Each line has been marked with a different small markers to indicate they are each of different lengths. The 3 angles inside will also be different as we will see later.

If just 2 sides are of equal length, it is called an **ISOSCELES TRIANGLE**:

Again, the two sides of equal length have been marked with identical markers. As for the two base angles a and b:
If we cut the triangle down the centre (mathematicians call this bisecting the triangle) we get:

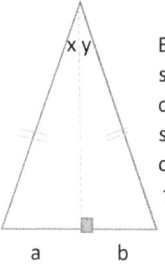

By doing this we see that the top angle is split in two. So angle x = y. Also the line crosses the base at right angles, making the smaller triangle on the left, identical to the one on the right as the side lengths are all the same. Hence the angles must also be the same. So angle a = b.

Hence for any ISOSCELES triangle, exactly two of the sides have the same length and the two angles from those sides are also equal.

If all 3 sides are equal it is called an **EQUILATERAL TRIANGLE**:

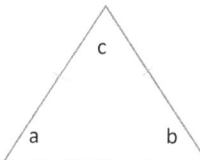

We indicate that the three sides are equal by marking each of them with an identical marker. As for the angles, a, b, and c, from above, if we bisect the triangle down the middle, we can see that angles a and b must be identical:

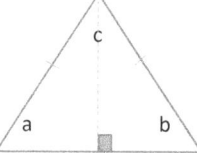

If we turn the triangle on its side and do the same, we can see that angles b and c are identical:

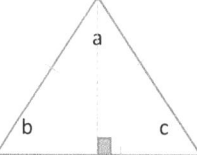

So if a = b and b = c, then a = b= c
And since all angles in any triangle add up to 180°, a = b = c = 60°
Hence for any EQUILATERAL TRIANGLE, all three sides are the same length AND all angles are the same and equal to 60°
NOTE: From above an equilateral triangle has 3 equal sides and 3 equal angles, an isosceles triangle has 2 equal sides and 2 equal angles. So the scalene triangle which has no equal sides, must have three different angles inside.

Examples

1. Find a,b,c

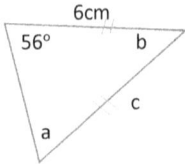

2. Find a,b,c

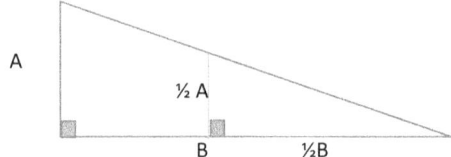

The markers indicate that it is isosceles,
So a = 56° and c = 6cm
Since all 3 angles must add up to 180°,
b = 180 – 56 – 56 = 68°

The markers indicate it is equilateral,
So a = c = 60°
and b = 4m

Similar Triangles

Similar triangles are ones in which the angles inside the triangles are identical and the sides are in the same ratio.

To prove this, can be a little tricky but here is a right angled triangle to show you the connection:

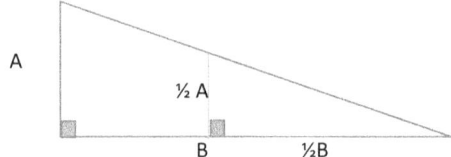

If we start with a right angled triangle, with shorter sides A & B, then create another right angled triangle inside it so that the two short sides have been halved, we get the diagram above.

From Pythagoras' Theorem, in the larger triangle, the longest side is $\sqrt{(a^2 + b^2)}$.
In the smaller triangle, it is a little more complex: $(½ a)^2 + (½ b)^2 = ?^2$
$(½ a)^2 = ¼ a^2$ and $(½ b)^2 = ¼ b^2$
So $(½ a)^2 + (½ b)^2 = ¼ a^2 + ¼ b^2 = ¼ (a^2 + b^2) = ?^2$
So ? = $\sqrt{\frac{1}{4}(a^2 + b^2)} = \sqrt{\frac{1}{4}} \times \sqrt{(a^2 + b^2)}$ and $\sqrt{\frac{1}{4}} = ½$
$? = ½ \sqrt{(a^2 + b^2)}$

Which means that if we halve the two shorter sides we also halve the third longest side. I.e. We could say that the **3 pairs of corresponding sides are in the same ratio** (One set is half the length of the other set).
This is true for ANY triangle.

What about the angles? From the diagram above, it is clear to see that two of the corresponding angles are identical. There are two right angles and the two triangles share the far right angle. So by the rules above, the third pair of corresponding angles MUST also be the same. This leads us to the definition of Similar Triangles:

If 2 triangles are SIMILAR, then each of the corresponding angles are the same and each of the corresponding sides are in the same ratio.

Example:

For the following pair of similar triangles, find a and b:

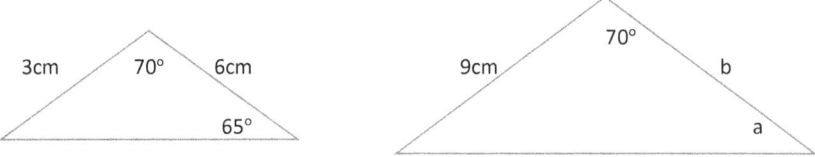

Sides are in the ratio: So 9/3 = b/6 -> b=18 (for ratio simply divide one side by the other)
Angles are the same, so a = 65°

Determining if two triangles are similar

What is the minimum amount of information that we need to say two triangles are similar?

1. From the definition above, if we knew that the three sides were all in the same ratio, then the three angles MUST be the same and the triangles are similar. We call this test, the side, side, side or **SSS test.**

2. Conversely if the three pairs of angles are the same, then the three sides MUST be in the same ratio and the triangles are similar. We call this test the angle, angle, angle or **AAA test**.

3. A third test states that if TWO of the sides are in the same ratio AND the included angles are identical then the triangles are similar. We call this test the side, angle, side or **SAS test**.

    For this third one, it is proven by seeing that you can only construct one triangle form two given sides and its included angle. Here is an example:

    From the diagram on the left, there is only 1 line that can be drawn to create a triangle. The same goes for the right diagram. Hence the two triangles MUST be similar.

4. Similar to test 3 above, a fourth test states that if two corresponding angles are the same and the included sides are in the ratio, then the triangles are similar. We call this test the angle, side, angle or **ASA test**.

    In the diagrams above, extending the two dashed sides at the angles shown, can only produce the one unique triangle. Hence the two triangles MUST be similar.

Examples:

Show that the following triangles are similar and hence find a:

1.

2cm	8cm
25° 109°	25° a
5cm	20cm

The two triangles are similar by the SAS rule. Hence a = 109°

2.

64° 4m		10m 64°	
3m 46°		46° a	
70°		70°	

The two triangles are similar by the AAA test. Hence sides must be in the same ratio.

NOTE: Which ratio is that? To get the ratio and hence the answer, divide one length by its corresponding other length. So to find a:

10/4 = a/3 -> 2.5 = a/3 -> 2.5 x 3 = a -> a = 7.5m

Teaching Points

1. The main emphasis here is to start with a simple fact (angle of a straight line is 180°) and build on it to form a range of more complex rules. This is a very useful application of deductive reasoning, which is a fundamental thinking tool in Maths.

2. For the Similar Triangles section, it may pay you to take some time out and explain more fully the concept of ratio. The main point here is simply that ratio is another way of doing divides. Also some knowledge of Pythagoras' Theorem would be handy.

3. Some students may find the proof of similar triangles a little difficult. You can alternatively, do it by experiment. By actually drawing specific triangles, you can show that each time you get similar triangles. For instance, draw a triangle. Now double each side and draw another triangle (also this will show that the third side must be fixed). Measure the angles inside of each and show they are identical.

Extensions

1. Why stop at triangles? You could extend the idea to quadrilaterals, pentagons, etc. There is a fundamental rule that connects the number of sides to the sum of the angles.

2. There are many drawing exercises that students may find interesting. Here are a few:

   a. **The Golden Triangle** – Draw an isosceles triangle with base length 12cm and base angles each of 72°. The ratio of base length 12 to length XY is called the golden ratio (about 8/5). The large triangle XYZ is called the Golden Triangle. Now draw a line from one base corner to the opposite side so that it bisects the angle (72°) as shown below:

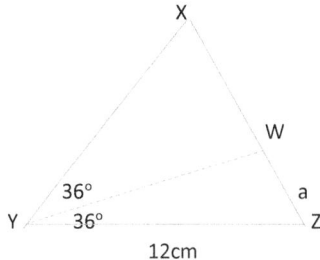

Also you will find that triangle YWZ is also a Golden Triangle. You can repeat the process by bisecting the angle at W to create yet another golden triangle and so on. Also starting at W, carefully cut out each side that is bisected you will get a neat spiral pattern.

b. Drawing polygons. Start with a circle of say 3cm diameter. If you want to say construct a 12 equal sided polygon (called a dodecahedron), you merely divide $360°$ by 12 (=$30°$). Then mark off every $30°$ on the circle and connect the dots. You will then have a dodecahedron. This can be done for ANY number of sides. What does a 20 sided shape look like? What happens to the shape as you increase the number of sides?

3. You could investigate similar patterns in 3D shapes. For instance, an investigation of the Platonic solids and why there are only 5 of them (check out Euler's formula). Students also enjoy constructing various 3D shapes. There are several instruction sheets and patterns available on the internet.

Algebra

Some practical examples

1. "$10 off everything in the store! ". If you wanted to calculate how much you wanted to pay you simply subtract 10 off the marked price. In Mathematics, we could write this down as a formula:

   p-10 where p represents the original price.

2. The number of points in football is the number of goal times 6 plus the number of behinds. As a formula, this could be:

   Gx6 + B (g=goals, B=behinds). (isn't that 'x' confusing! Is it x or times?)

To make thing less confusing, we drop the 'x' and always put the number in front. So it looks like this:

   6G + B

This is what Algebra is all about. We use letters of the alphabet (and other symbols) to represent numbers. The letters are known in Maths as **Pronumerals.** The 'formulas' above are sometimes called **algebraic expressions** and the individual parts (e.g. '6G') are known as **terms**.

Combining these terms can be tricky, so here is an alternative way of looking at it....

Groups

I use the idea of groups instead of terms. Each group is known by the letter or combination of letters in the term

E.G. a, 5a, -6a all belong to the group of 'a''s

   xy, 2xy, -9xy, 3yx all belong to the group of 'xy''s (Note the order of the letters does not matter)

Same and Similar Groups

Exercise: Pick out which terms belong to the same group in the following list:

7P, 8Q, 6P, 9PQ, -4P, -3QP

Answer: 7P, 6P and -4P are all in the same group. 9PQ and -3QP are in the **same group**.

But 7P and 8Q are in **different groups**.

7P and 9PQ are also in different groups. But, hang on! They both have 'P' in them. In these cases we say that they are in **similar groups**. This is also the case when we introduce indices into the mix.

E.G. Here is a list of Algebraic terms:

   $3a, 5b, 4abc, -7bac, 2a^2, 3ab^2, 5a^2b$

3a and 5b are in **different groups** (No common letters)

4abc and -7bac are in the **same group** (both sets of letters are identical)

3a, -7bac are in **similar groups** (the 'a' is common to both)

$3a, 2a^2, 3ab^2, 5a^2b$ are all in **similar groups** (again the 'a' is common to all)

Let's start with some simple examples:

$3a + 2a = 5a$ Why? Well, even though we don't know what 'a' stands for, we do know that 3 of them and 2 of them will add to 5 of them.

$3a + 2b$ = ? As we don't know what 'a' stands for and what 'b' stands for, we can't add them. So **$3a + 2b = 3a + 2b$**

$3a + 2a^2$ = ? This is the tricky one. Most students will attempt to combine the two terms somehow. But as a^2 is really a x a, and we don't know what 'a' stands for we can't really combine them.

So **$3a + 2a^2 = 3a + 2a^2$**

This leads us to a simple rule:

You can only add and subtract terms which belong to the SAME group

---

Terms which belong to **similar** or **different** groups cannot be added or subtracted and thus you simply write out the problem again. Here are a few worked examples:

1. $9x - 7x = 2x$ (same groups)
2. $6p + 9q = 6p + 9q$ (similar groups)
3. $4ab - 3a = 4ab - 3a$ (similar groups)
4. $12x^2y + 6xy^2 = 12x^2y + 6xy^2$ (similar groups)

This rule can then lead onto simplifying algebraic expressions:

E.g. Simply $x^2y + 6xy - 2x^2y + 3xy + 5x$

$= - x^2y + 9xy + 5x$ (The 'x^2y' terms combined, the 'xy' terms combined, the 'x' doesn't, so it is left on the end)

Notes:

1. When terms in the same group are combined, it is only the numbers in front which are added or subtracted ($2a + 3a = 5a$; $2 + 3 = 5$)
2. When two terms in the same group are combined, the answer is also in the same group (from above, 2a, 3a and 5a are all in the same group). The only exception is when the two terms combine to give '0' as the number in front. E.g. $-3a + 3a = 03a$ which is 0.
3. If there is no number in front of the term, it is actually meant to be a '1'. ($3a + a$ really should be $3a + 1a = 4a$). It is a good idea for weaker students to get in the habit of putting the '1' in there. This also applies for subtraction. E.g. $3a - a = 3a - 1a = 2a$
4. What about numbers on their own? The simple rule is that **numbers form their own group**. E.g. $7x + 5 = 7x + 5$ because 7x and 5 are different groups

Here are a few final examples. Simplify the following:

1. $3x + 5y - 7x + y = 10x + 6y$
2. $2p + 9pq - 6p + 9 = -4p + 9pq + 9$ (could also be written as $9pq - 4p + 9$)
3. $4ab - 2a + 8ab - 2a + b = 12ab - 2a + b$
4. $5x^2yp + 6xy^2 + 2px^2y - 6xy^2 = 7x^2yp$

Two Special Cases:

1. Single terms

E.g. $3a + a = ?$

The two terms are in the same group, so can be added. In reality, it is really asking, what is 3 a's plus another a. Answer is 4a.

$3a + a = 4a$

Hence the single 'a' should have been written as 1a : $3a + 1a = 4a$

This leads us to another simple rule:

Single terms, such as a should have a 1 in front, i.e. 1a

E.g. Simplify the following:

1. $4ab + a + 3ab + 2a + b = 7ab + 3a + b$
2. $5x^2yp + xy^2 + 4px^2y - 6xy^2 = 9x^2yp - 5xy^2$

2. Plus and Minus Signs

E.g. $3a + 2b - 5a + b$. Another simple rule to follow to avoid confusion about signs is simply that any sign in the expression attaches itself to the term or number that follows:

So from above, the first + is attached to the 2 which makes that '2' positive. Similarly the 5 is negative because it has a minus sign in front of it. The '+b' should be '+1b' which makes the 1 positive 1. Finally, the '3a' has no sign in front of it but when calculating the answer, you will combine $3a - 5a$. Hence the 3 is regarded as positive. So if there is no sign in front of the term, then it is positive.

(While this may not be overly relevant in adding and subtracting terms, it becomes quite important when multiplying and dividing terms and when you get to solving linear equations)

Multiplication and Division of Algebraic Expressions

Terms in the same group

E.g. $2a \times 3a = ?$

Recall that 2a really means 2 x a and 3a is really 3 x a.

The expression should be therefore written as

$2a \times 3a = 2 \times a \times 3 \times a = 2 \times 3 \times a \times a$ (in multiply, order does not matter)

$= 6 \times a^2 = 6a^2$

So, $2a \times 3a = 6a^2$. We have multiplied the numbers (recall numbers form their own group), and we have multiplied the 'a's using basic index laws. I.e. we have multiplied each part individually.

A few further examples:

$2b \times 4b^2 = 2 \times 4 \times b \times b^2 = 8 \times b^3 = 8b^3$ (note that 2b should be written as $2b^1$)

$6ay^2 \times -2ay^2 = -12a^2y^4$

$pa^2y \times 2a^2yp = 2p^2a^4y^2$ (from before pa^2y should be written as $1pa^2y$)

Terms in different groups
E.g. $3a \times 4b = 3 \times a \times 4 \times b = 3 \times 4 \times a \times b$

Well $3 \times 4 = 12$, that's obvious but since we don't know what 'a' and 'b' are we really can't combine them. So,

$3a \times 4b = 12ab$

Note that we still multiply individually but terms (or parts of terms) in different groups won't combine

Some more examples

$6m \times 2n = 6 \times 2 \times m \times n = 12mn$

$3a^2 \times 2b = 6a^2b$

$5c^3 \times d = 5c^3d$

Terms in Similar Groups
From basic index laws if we have say $a^3 \times a^4 = a^7$. So terms in similar groups can combine.

E.g. $5b^2 \times 4b^2 = 5 \times 4 \times b^2 \times b^2 = 20 \times b^4 = 20b^4$

So, once again you can combine terms in similar groups and you multiply individually

Some examples:

$4c^2 \times 2c^3 = 8c^5$

$3ab^2 \times 2abc = 6a^2b^3c$ (In complex examples like these, the method is to match the parts that are in the same of similar groups, i.e. 3×2 – don't forget, numbers form their own group, $a \times a$, b2 x b and multiply individually. Any parts, that don't pair up or are in different groups, are simply tacked on the end)

$-6bc \times cd = -6c^2bd$

What about divide??
Since divide is the opposite of multiply, then the same rules must apply. In algebra, we write down divides as fractions. (Fractions are divides and divides are fractions)

Here are some basic examples to show you how it works

$\frac{3a}{6a} = \frac{3 \times a}{6 \times a} = \frac{3}{6} \times \frac{a}{a} = \frac{1}{2}$ (In this example, it is easier to leave the divides as fractions since 3/6 cancels down to ½ and a/a simply cancels out)

$\dfrac{8a^2}{4a} = \dfrac{8 \times a^2}{4 \times a} = \dfrac{8}{4} \times \dfrac{a^2}{a} = (8 \div 4) \times (a^2 \div a^1) = 2a$

$\dfrac{12a^3b}{3ac} = \dfrac{12 \times a^3 \times b}{3 \quad a \quad c} = \dfrac{4 \times a^2 \times b}{c} = 4a^2\dfrac{b}{c}$ (In this example, the b and c don't match so are left alone)

So, the basic rule is:

> ## You can Multiply and Divide terms in the same and similar groups only

Notes:

1. The method is to match parts which belong to the same or similar groups and multiply or divide individually.
2. When dividing, you tend to cancel as a fraction as most divides in algebra are written as fractions
3. When multiplying, you add indices and when you divide you subtract or cancel indices. If there is no index, put in a 1

Some final examples

$8a^2 \times 2ac = 16a^3c$

$3ba^3 \times -ab^3 = -3b^4a^4$ (in this example, the minus is actually -1)

$\dfrac{2ab}{8a} = \dfrac{b}{4}$

$\dfrac{10a^2}{4a^5} = \dfrac{5}{2a^3}$ (Another way of looking at the indices in this example is to cancel '2' out of the top and bottom in the 'a' terms leaving a^3 on the bottom)

Teaching Points

1. Most students make basic mistakes with algebra, mainly in trying to combine all the terms into one single term no matter what. One classic example is to get your students to simplify a+3. Most will come up with 3a.
2. The intention of the method described above is twofold. Firstly it makes clear to students that there sometimes is no simplification to problems (a+3 = a+3). The second intention is to provide a simple set of rules to solve even the most difficult of algebra problems. Try this one:
   $4a^2 + 3ba^2 - a^2b + 2a^2 - a = 6a^2 + 2ba^2 - a$
3. This method should be taught AFTER a good solid look at indices. Also having an understanding that **divides are fractions** would be helpful. This method really reinforces these concepts.
4. One of the issues you have with students and algebra is getting them to see the relevance of it. One approach to this is to link real world problems to algebra. There are

many books, problems and puzzles devoted to this. I've listed a few in the Extension activities below.

Extension Activities

1. **Brackets in Algebra**

    Like basic terms, any expression in brackets will also form its own group.

Example

Simplify $8(x+2) - 3(x+2)$

The $(x+2)$ term forms its own group. So the answer is $5(x+2)$

Notes:

1. For an expression like $(2x-3)^2 + 4(2x-3) + 5$ can't be simplified as $(2x-3)^2$ and $4(2x-3)$ are similar groups not the same.
2. To further simplify brackets, you really need to eliminate them. See the section on expanding and factorization for this.

2. **Modelling Maths**

    To get a better appreciation of the role Algebra plays in the real world, here are a few problem solving activities, you might try. For each of them, the path to the solution lies in being able to convert the pattern you find into an algebraic expression.

1. The Tennis tournament
    Your task is to organise a knockout tennis tournament for 50 contestants. The two main problems are:-
    (a) How many games would a contestant have to play to win the competition? Could you express this as a formula given any number of contestants? Also how many games would be played in total?
    (b) Assuming that there is a total of $500,000 in prize money to be divided up amongst the contestants, so that every contestant gets something, how would you divide up the $500,000?

    Other things you may want to investigate:

    (a) What if you had only four tennis courts? How would you organise the competition?
    (b) What if the competition was a round robin where every contestant plays every other one, how many games would be played then?
2. Population growth
    If we consider a simplified version of this, and assume that each family has four children, each parent will have two offspring (on average).
    Hence, for a parent, the first generation would produce 2 children. They, will then produce two children each for the second generation (a total of 4 children now). Each of these four will then produce two children each for the third generation giving a total of 8 children. Is there a way of predicting how many children will be produced for a given generation?

Other things to consider. What would happen if each parent produced on 1 child or say 1.5 children (on average). Can you adjust your formula to consider how many children will be produced for a given generation? (Hint: It may pay you to graph the results first)

3. The famous Tower of Hanoi Puzzle
   This famous puzzle has a board and 3 pegs on it, with 8 disks stacked on one peg (shown below). The challenge is to transfer the stack of 8 disks to one of the other pegs in the fewest possible moves, moving 1 disk at a time and NEVER placing a disk on top of a smaller one. How many moves will it take for 8 disks?

How many moves will it take for say 6 disks? How about 4 disks? The next challenge is to see if you can find a formula that tells you the minimum number of moves given the initial number of disks?

Brackets and Factorisation

These topics can prove to be difficult for some students. The crucial aspect is that they need to be able to find the highest common factor between 2 numbers. This may take some practice.

Brackets

Recall from Algebra, that a term like 3a belongs to the group of 'a's. Well brackets form their own groups as well. For instance 3(x+6) belongs to the group of (x+6).

Hence the same Algebra rules apply:

You can only add and subtract terms in the SAME group

E.g. $4(a-2) + 2(a-2) -3(a+2) = 6(a-2) -3(a+2)$

You can only multiply and divide terms in the SAME or SIMILAR groups

E.g $3(a+4) \times 4(a+4)^2 = 12(a+4)^3$

NOTE: What about the terms 2(a+2) and 3(a+3). Are they in SIMILAR groups? They both contain 'a'. As you know, if we were to multiply them and remove the brackets, ultimately we would end up with an 'a^2' term. So they kind of combine as similar groups. However, for simplicity, we regard them as two different groups while they remain in brackets.

Expanding Brackets

While mathematicians regard brackets as elegant, most students regard them as nasty. We even had brackets as one of the 'nasties' in our linear equation formula. So the process of expanding brackets is common and it allows us to split the terms inside the brackets and get on with the business of simplifying.

The basic rule, and indeed what brackets mean is:

The term outside the brackets is MULTIPLIED by EACH term inside the brackets.

E.g Expand 5(y + 9)

$5(y+9) = 5xy + 5x9 = 5y +45$

E.g. Simplify the following by expanding the brackets:

$2(a + 3) - 3(2a + 4)$

$2(a + 3) - 3(2a + 4) = 2a + 2\times3 -3\times2a -3\times4 = 2a + 6 -6a - 12 = -4a -6$

NOTE: the minus sign in the example above is significant. Remember, any minus sign in an expression attaches to the term that follows and is considered to be negative. So '-3(2a+4)' means to multiply the brackets by -3.

E.g Simplify the following by expanding the brackets:

1. $4(3a - 2b) - 2(a - b) = 4\times3a - 4\times2b -2\times a -2\times-b = 12a - 8b - 2a + 2b = 10a - 6b$

2. $5(2a + 3b) - (a - b) = 5\times2a + 5\times3b - 1\times a -1\times-b = 10a + 15b - a + b = 9a + 16b$

NOTE: In the second example, there is nothing in front of the brackets (except -), so it is actually 1. The problem should have read as 5(2a + 3b) – 1(a – b) but the '1' is usually left out.

Factorising (putting into brackets)

First step in factorising is revise finding the highest common factor. That is, the largest number that divides exactly into 2 other numbers. There is no real shortrcut to this, except being familiar with your times table.

E.g. Find the highest number that will divide exactly into 36 and 42.

Answer is 6. The only real trick here is to think of numbers that divide into say 42 (there is only 6,7, 2, 21). From these the highest that also goes into 36 is 6.

To factorise, the first step is to find the highest common factor of the two terms. This is the term outside the brackets. The terms inside are found by dividing by the outside term

E.g. Factorise 12a – 8b

Highest number that goes into 12 and 8 is 4. Dividing each term by 4, we eventually get:

$12a - 8b = 4(3a - 2b)$

This is all well and good for numbers but what about pronumerals? For instance:

E.g. Factorise fully $8a^2b + 6ab$. To fully factorise, we pair off the parts that belong to the SAME or SIMILAR groups. I.e Pair off the numbers 8 & 6, a^2 & a, and b & b. Find the highest common factor of each, then divide that into each part.

We should see that the highest number that goes into 8 and 6 is 2. But what about a^2 and a? We don't know what 'a' represents but we do know that a^2 is a x a. So that highest term that divides exactly into a^2 and a must be a. Also for b and b the highest term would be b, even though we don't know what 'b' is.

So, $8a^2b + 2ab = 2ab(2a + 3)$

This leads us to a simple set of steps to factorise:

Step 1: Pair off the parts that belong to the SAME or SIMILAR groups

Step 2: Find the highest common factor of each pair. For terms, it is the one with the lower power. This is the term outside the brackets.

Step 3: Divide each term by the highest factor for the terms inside the brackets

Examples

Factorise fully each of the following

1. $2a^3b^2c + 10a^2b^4 = ?$

Step 1: 2,10 a^3,a^2 b^2,b^4 . 'c' has nothing to match with

Step 2: Highest common factor: 2,10 = 2 $a^3,a^2 = a^2$ $b^4,b^4 = b^2$.

Step 3: $2÷2 =1$, $10÷2=5$, $a^3÷a^2 = a$, $a^2 ÷a^2 =1$ $b^2÷ b^2= 1$, $b^4 ÷b^2= b^2$.

So the answer is: $2a^3b^2c + 10a^2b^4 = 2a^2b^2(ac + 5b^2)$ NOTE: The 'c' is left alone

2. $4a^4b^2 - 6a^5b^4 = ?$ A slightly quicker way is to use arrows to pair off and find the highest common factor

$4a^4b^2 - 6a^5b^4 = 2a^4b^2(2 - 3ab^2) = 2a^4b^2(2 - 3ab^2)$

3. $5a^2b - 15ab$

$15a^2b - 5ab = 5ab(3a - 1) = 5ab(3a - 1)$ Note: in this case the second term completes divides by the

Highest common factor, so the result is 1 as anything divided by itself is 1.

4. $4a^3b^2 - 8a^5b^4 - 12a^2bc$

$4a^3b^2 - 8a^5b^4 - 12a^2bc = 4a^2b(ab - 2a^3b^3 - 3c)$ Yes it can even be done with 3 or more terms.

Teaching Points

1. The main point is to emphasise that expansion means to multiply ALL terms inside the brackets.
2. Brackets form their own group which means that the rules of Algebra also apply
3. Students find factorising difficult for some reason. If you break it down into two separate concepts – Highest Common Factor for numbers & Highest Common Factor for powers – and show that each is quite simple and straightforward, it becomes a much easier exercise.

Extension work

1. Expansion and factorisation work no matter how many terms you have in the brackets. See example 4 above. You could try even more terms and show students that they can master some very difficult expressions.
2. Highest Common Factor of fractions. You could extend the idea to include fractions. It won't take much to show that the H.C.F. of denominators is really the same as for ordinary numbers and powers!

Linear Equations

These are equations where there is an unknown value (usually 'x') which represents some missing number.

An example:

$3x + 4 = 10$ (3x means 3 times x. Using the 'x' symbol for times would be just confusing!)

The aim is to find out what missing number is x? That is, what number can be multiplied by 3 then add 4 and give an answer of 10? You could guess the answer by simply trying various alternatives. In this case that might work but what we want, is a guaranteed process that gives you the answer each time.

Here is alternative way of approaching it:

Let's start with a very simple example:

$x + 4 = 7$

The aim of our method is to do something to this equation so it reads **x = the answer**. I.e. get 'x' by itself on one side of the equation.

The classic approach is to do the following:

Subtract 4 from both sides. Why? Because the equation should be equal as we are removing the same amount from both sides. Why subtract 4? Simple! We can see a step ahead, that if we subtract 4 then the + 4 on the left disappears leaving x by itself:

$x + 4 - \mathbf{4} = 7 - \mathbf{4}$

$x = 3$. We have the answer.

Now an alternative approach:

Look at the equation above and see what happened to the 4. It went from one side of the equation to the other and changed its operation to its opposite (plus to minus).

$x + 4 = 7 - 4$

$x = 3$

Let's try another example:

$x - 5 = 8$

Using this new strategy, we firstly aim to get 'x' by itself. To do this, get rid of the '-5'. To do this, shift it to the other side and change to the opposite operation (i.e. +5)

$x - 5 = 8 + 5$

$x = 13$

Does this work with multiply and divide?

E.g.   4x = 12. To get 'x' by itself, get rid of the 4 (which is multiplied by the 'x'). Shift it to the other side and change to the opposite operation (i.e. ÷ 4)

4x = 12 ÷ 4

x = 3. The right answer (4 x 3 = 12)

E.g. $\frac{x}{3}$ = 2. Get 'x' by itself. Get rid of the 3 (which is dividing into the 'x').
        Shift it to other side and change it to its opposite operation (i.e. x 3)

$\frac{x}{3}$ = 2 x 3

x = 6. Again the right answer (6/3 = 2)

Note: In linear equations, multiplying unknowns is shown by the number **IN FRONT OF** the unknown and we drop the times symbol. For instance, a x 6 is written as 6a not a6. Similarly, we represent all divides as fractions. So for instance a ÷ 4 is written as a/4)

So for simple equations our method so far is:

Aim: to get the unknown by itself.
Method: get rid of the numbers around the unknown by shifting them to the other side change to the opposite operation

Some examples:

a – 7 = 10   a = 10 + 7   a = 17

9a = 27   9a = 27/9   a = 3

More than one number with the unknown

E.g.   2a + 3 = 11

Using our method above, we aim to get the 'a' by itself by removing the '2' and '3'. Also we can see that the '2' is multiplying the 'a' and the 3 is added to the 'a'.

We then proceed to place the number on the opposite side with their opposite operation. But which one should go first? Does it matter?

The easiest way to answer these is to test it out:

(a) Let's get rid of the '3' first then the '2' –
      2a + 3 = 11/2
      a + 3 = 5.5 – 3
      a = 2.5
(b) Now get rid of the '3' first then the 2
      2a + 3 = 11-3
      2a = 8 / 2
      a = 4

There can only be one right answer. The simple check is to replace 'a' by each answer and see which one works:

(a) $2 \times 2.5 + 3 = 8$ not 11
(b) $2 \times 4 + 3 = 11$

So method (b) gives the right answer, a=4. But why does the second way work and not the first?

The answer to this relates back to something you did a long time ago in Arithmetic.

2a + 3 really means 2 x a THEN add 3.

A little history lesson:

Remember in arithmetic, problems like this 7 + 3 x 4. The order you calculated each part was important. It was 3 x 4 first then add to the 7. You even had an acronym to help you remember. It was BODMAS – Brackets, Of, Multiply, Divide, Add, Subtract.

Returning to our equation above, 2a + 3 = 11, the correct sequence we removed each number was to get rid of the Add then the Multiply. I.e. the opposite order to BODMAS! It make sense, since the equation really says 2 x a THEN add 3.

A few examples:

1. 4a - 9 = 3   (4 is times and 9 is subtract – get rid of the 9 first then the 4)
    4a -9 = 3+9 = 12
    4a = 12/4
    a = 3

2. a/5 + 4 = 7   (5 is divide and 4 is add – get rid of the 4 first then the 5)
    a/5 + 4 = 7-4 = 3
    a/5 = 3 x 5
    a = 15

    So we can modify our method thus:

Aim: to get the unknown by itself.

Method: get rid of the numbers around the unknown by shifting them to the other side change to the opposite operation.

The order of elimination is BODMAS backwards (or SAMDOB if you like)

Some further examples: (check out each to make sure we have the right answer)

1. 8a - 4 = 12   (8 is times and 4 is subtract – get rid of the 4 first then the 8 - SAMDOB)
    8a -4 = 12+4 = 16
    8a = 16/8
    a = 2

2. 3a/2 + 5 = 11   (3 is multiply, 2 is divide and 5 is add – by SAMDOB, the order of elimination is 5 then 3 then 2. – Actually you could do 5 then 2 then 3. It's just that SAMDOB sounds better than SADMOB. Well maybe not!)

$3a/2 + 5 = 11 - 5 = 6$

$3a/2 = 6/3 = 2$

$a/2 = 2 \times 2$

$a = 4$

3. $4(a - 2) + 7 = 19$

Careful with this one. I've introduced brackets. But it should still work. The '4' is multiplying the 'a', even though there is no multiply sign and it is outside the brackets. The '2' is being subtracted BUT, being inside the brackets, it is treated as a bracket operation. The '7' is being added. So according to SAMDOB, the order of elimination should be 7 then the 4 and finally the '2'.

$4(a - 2) + 7 = 19 - 7 = 12$

$4(a - 2) = 12 / 4$

$a - 2 = 3$ (The brackets can be dropped now as there is nothing outside the brackets)

$a - 2 = 3 + 2$

$a = 5$

"Nasties"

The above method should let you easily solve most linear equations. But there some more difficult types for which the method may not directly apply. I refer to these equations as "Nasties". There are 4 basic types of Nasties.

1. **The Hidden brackets**

Brackets are usually left out of the expression when you have divides. Here is an example:

$\dfrac{a + 3}{6}$ The expression on top should have brackets around it. The reason is simply that the bottom number is dividing ALL of the top expression.

So, this expression should be:

$\dfrac{(a + 3)}{6}$ Hence, when eliminating the numbers, SAMDOB tells you now that the 6 goes first then the 3

An example

$\dfrac{a - 4}{3} + 8 = 9$ The nasty here is the hidden bracket. So put it back in.

$\dfrac{(a - 4)}{3} + 8 = 9$ So now the order of elimination will be the '8', then the '3' and then the '4' By SAMDOB, brackets are done last.

$\dfrac{(a - 4)}{3} + 8 = 9 - 8 = 1$

$\dfrac{(a - 4)}{3} = 1 \times 3 = 3$

$a - 4 = 3 + 4 = 7$ (Can drop the backets now – nothing outside them)

$a = 7$

2. **More than one unknown**

This simply means that the unknown appears at least twice in the equation.

E.g. $3a + 5 = 6a - 4$

To solve these, we would prefer if we had only one unknown. So our first step is to change the equation so there is only one unknown. We do this by using the same method as above. If we treat, say the '3a' as a single term, then it can be shifted to the other side but must be done in SAMDOB order. As the 3a is really an add, we can shift it straight away. So:

$3a + 5 = 6a - 4 - 3a = 6a - 3a - 4 = 3a - 4$ (we can then combine the two 'a' terms by mere subtraction)

$5 = 3a - 4$ (Now we have a standard equation like above. We can remove 4 first then the 3)

$5 + 4 = 3a - 4$

$9 = 3a$

$9/3 = 3a$

$3 = a$

NOTE: we could have shifted the '6a' instead but you might predict that you would be dealing with negative terms. Not a pleasant prospect. However, it would be a good exercise to try shifting the '6a' instead and see if you get the same answer. The hint here is that it is better to shift the smaller term rather than the larger term. Also note that the unknown ends up on the right hand side, not the left. This again, does not matter. All we want is the unknown by itself on one side of the equation. Which side, doesn't matter.

So to eliminate this nasty, get the unknowns together on the same side by shifting one of them as a single term but you must do it in SAMDOB order. Then simply combine them into one term.

One more example, just to prove it works:

$5 - 2a = 5a - 9$ Identify the nasty. 2 unknowns. Shift one to the other side. Better to shift the smaller one. So shift '2a' to the right.

$5 - 2a = 5a - 9 + 2a = 5a + 2a - 9 = 7a - 9$

$5 = 7a - 9$ Now get rid of the 9 then the 7 by SAMDOB

$5 + 9 = 7a - 9$

$14/7 = 7a$

$2 = a$

3. **Unknown on the bottom**

E.g. $\dfrac{12}{a} + 7 = 11$

In these cases, we really want the 'a' on top so we can solve the problem. Hence the solution to this nasty is to get the 'a' on top. The 'a' is really being divided, so we could shift it to the other side, **but we have to keep the SAMDOB order.**

So, for our example, the method would be to shift the '7' first as it is an add, then shift the 'a' as a divide.

$\dfrac{12}{a} + 7 = 11 - 7 = 4$

$\underline{12} = 4 \times a = 4a$
(a)

$12 = 4a$ (now simply shift the 4 across to leave 'a' on its own)

$\underline{12} = 4a$
4

$3 = a$

One more example, just to prove it works:

$\underline{12}$ - 5 = 1 Identify the nasty. Unknown on the bottom. Get rid of this by shifting the
a   'a' but must shift the '5' first as it is subtract and the 'a' is divide

$\underline{12}$ -5 = 1 + 5 = 6
a

$\underline{12}$ = 6 x a
(a)

$\underline{12} = 6a$ Nasty is now gone. Solve it like an ordinary equation. Shift the 6 across
$\underline{12} = 6a$
6

$2 = a$

These last two nasties are really optional ones as there are other ways you can eliminate them.

4. **Minus signs and the missing 1**

e.g. $4 - a = 6$

There really should be a 1 in front of the 'a'. So it should be '-1a'.

$4 - 1a = 6$

Now to solve this, simply shift the numbers in order. From our algebra work, we learnt that any '-' sign attaches to the thing that follows and makes it negative. So the '1' we added is really '-1' and it is multiplying the 'a'. What about the '4'? It must be an add. Why? Well. The '4' has no sign in front of it. So we can put a '+' in front as the number is really positive 4.

$+4 - 1a = 6$

We could re order the expression thus (Note the signs stay with the numbers that follow)

$-1a + 4 = 6$ In this form you can see that the '4' is added so it is shifted first and becomes minus.

$4 - 1a = 6 - 4$

$-1a = 2$ (Note that we have not touched the -1, so it stays where it is. Now shift the '-1')

$-1a = 2 \div -1 = -2$

$a = -2$

As an alternative, you could have solved this equation thus:

$4 - a = 6$ (The nasty is the –a so shift it. As it is –a we can shift it to the other side and make it +a

$4 - a = 6 + a$

$4 = 6 + a$ (now get rid of the 6)

$4 - 6 = 6 + a$

$-2 = a$

NOTE: From our algebra, numbers or terms with no sign in front of them, are positive and can be interpreted as added. E.g. 9 is +9, 2a is +2a. Similarly negative signs with NO number following are really -1. E.g. –a is really -1a and is interpreted as -1 x a

5. **Brackets**

E.g $4(2a + 3) = 36$

Most students don't handle brackets very well. So get rid of them. Recall that anything outside the brackets is multiplied by EACH term inside the brackets:

$4(2a + 3) = 36$

$4 \times 2a + 4 \times 3 = 36$

$8a + 12 = 36$ Now it is just an ordinary equation. Solve it by getting rid of the +12 first then the 8

$8a + 12 = 36 - 12 = 24$

$8a = \dfrac{24}{8}$

$a = 3$ (as a check, $4(2 \times 3 + 3) = 4(6 + 3) = 4 \times 9 = 36$)

These 'nasties' then modify our basic rule to:

Aim: to get the unknown by itself.

Method:

- **get rid of any 'nasties' first. (unknown more than once, unknown on the bottom, brackets etc.). Then**

- **get rid of the numbers around the unknown by shifting them to the other side change to the opposite operation.**

The order of elimination is BODMAS backwards (or SAMDOB if you like)

Some worked examples:

1. $\dfrac{a - 5}{6} + 2 = 7$

    Step 1: Get rid of any 'nasties'. There should be brackets around the a-5.

    $\dfrac{(a - 5)}{6} + 2 = 7$ Step 2: eliminate numbers around 'a' in order. From SAMDOB, the order is the +2 first, then the ÷6 and finally -5 as it is in brackets.

    $\dfrac{(a - 5)}{6} + 2 = 7 - 2 = 5$

    $\dfrac{(a - 5)}{6} = 5 \times 6$

    $a - 5 = 30 + 5$ (the brackets can be dropped now as nothing is outside them)

    $a = 35$

2. $2 - \dfrac{7}{3a} = 12$ Step 1: Get rid of any nasties. Unknown on the bottom. Shift to the other side and multiply. BUT must do it in order. So shift the '2' first.

$2 - \dfrac{7}{3a} = 12 \rightarrow 2 = 10$ (note that the '-' sign remains attached to the 7 and the 2 is plus)

$-\dfrac{7}{3a} = 10 \times 3a = 30a$ (you could have shifted just the 'a' if you wanted)

$-\dfrac{7}{30} = 30a$

$-\dfrac{7}{30} = a$ (not all answers are nice whole numbers!)

Teaching Points

- This method provides a basic rule that should solve nearly all linear equations. Students should be able to summarize the method into a simple set of rules.
- It should be taken slowly with many examples to reinforce the concepts.
- It also reinforces the basic Algebra ideals of attaching signs to the term that follows and brackets
- The last examples above are probably about as hard as you are ever likely to get in linear equations.
- It should also be emphasised that once you have an answer, it can be substituted back into the original equation. If you get the left hand side = right hand side you have the correct answer. It is a rare case where you know you have the right answer without having to look up the answers.
  Example:
  Solve $3a - 4 = 11$
  $\qquad 3a = 11 + 4 = 15$
  $\qquad a = 15/3 = 5$
  Sub a = 5 back into the equation:
  $3 \times 5 - 4 = 11$
  $15 - 4 = 11$
  $11 = 11$. So a = 5 MUST be the right answer.

Extension Activities

An intro to quadratic equations

Example:

$3a^2 + 5 = 17$

Using the same method as above, you would eliminate the numbers around the unknown in order of SAMDOB. But where does the (^{2}) fit in?

In short, going back to the original order of operations, looking at just the left hand side of the example above, squaring is done BEFORE the Multiply and Add (I.e. You would square 'a' first then multiply by 3 then add 5).

For brackets, consider the example:

$(a + 4)^2 - 3 = 61$

Again for order of operations, you would be doing the brackets first, then the square, then then subtract for the left hand side. This leads us to include the square operation (and indeed its opposite, the square root) to fit in to the order of operations just after brackets. So the order then is:- BSSODMAS (the two S's stand for Square and Square Root).

So for order of elimination, we have SAMDOSSB!

So for: $3a^2 + 5 = 17$ - Get rid of the 5 first

$3a^2 + 5 = 17 - 5$ =12 Get rid of the 3 next

$3a^2 = 12/3$ =4 Get rid of the square last

$a^2 = \sqrt{4}$ =±2

a = ±2

And for $(a + 4)^2 - 3 = 61$ – Get rid of the 3 then the square, then the brackets (i.e. the +4)

$(a + 4)^2 - 3 = 61 + 3$ =64

$(a + 4)^2 = \sqrt{64}$ =±8

$a + 4 = ±8$ -4 = 4 or -12

Surds

Surds are actually $\sqrt{\ }$ of any number. They also provide a good approach to the concept of irrational numbers. Consider the solution to $\sqrt{2}$.

From a calculator $\sqrt{2}$ = 1.4142135623730950488016887242097....

However, the decimal places continue without end and NEVER repeat. This is an example of an Irrational number. The point to emphasise here is that the best way to write down $\sqrt{2}$ is $\sqrt{2}$!

These are called surds. Note that there are other numbers which are not surds which are Irrational. π for example.

NOTE: In fact the $\sqrt{\ }$ of ANY prime number is written as a surd.

The main arithmetic of surds is simplifying surds. There are close parallels with simplifying algebraic expressions

How does it work?

Consider a simple problem – find $\sqrt{36}$.

Now we know the answer is 6 (or ±6 to be precise) but we can use this to show a fundamental rule for surds. We can split the 36 up into 9 x 4, thus:

$\sqrt{36} = \sqrt{(9x\ 4)}$

But $\sqrt{36} = 6 = 3 \times 2$. So

$\sqrt{(9x\ 4)} = 3\ x\ 2 = \sqrt{9}x\sqrt{4}$

This gives us a fundamental rule for surds:

$$\boxed{\sqrt{(ax\ b)} = \sqrt{a}x\sqrt{b}}$$

Examples:

Simply

1. $\sqrt{48}$

$= \sqrt{16\ x\ 3}$ (The trick here is to split the 48 up into products that include a perfect square)

$= \sqrt{16}\ x\sqrt{3}$

$= 4\sqrt{3}$ (Leave the solution in surd form, similar to what you do in Algebra)

2. $\sqrt{3} \times \sqrt{6}$

$= \sqrt{3\ x\ 6} = \sqrt{18}$

$= \sqrt{9\ x2} = \sqrt{9}x\sqrt{2} = 3\sqrt{2}$

3. $6\sqrt{5} - 4\sqrt{5}$ (here is where the comparison to Algebra is very apparent. Just like algebraic terms, $\sqrt{5}$ can be treated as term in its own group. In Algebra, you can add and subtract terms in the SAME group, so the answer is..
$= 2\sqrt{5}$

4. $\dfrac{\sqrt{96}}{\sqrt{12}}$ Since ÷ is the opposite of x, the same rule applies, i.e.

$$\frac{\sqrt{a}}{\sqrt{b}} = \sqrt{\frac{a}{b}}$$

So:

$$\frac{\sqrt{96}}{\sqrt{12}} = \sqrt{\frac{96}{12}} = \sqrt{8} = 2\sqrt{2}$$

A special case:

Rationalising the denominator

This simply means that if you have a $\sqrt{}$ on the bottom of a fraction, then we would prefer it was a nice whole number so we can actually divide. The trick in each case is to simply multiply top and bottom by that same $\sqrt{}$. This will immediately make the bottom number whole.

Example

Rationalise the denominator of the following:

1. $\dfrac{1}{2\sqrt{3}}$

   $\dfrac{1}{2\sqrt{3}} \times \dfrac{\sqrt{3}}{\sqrt{3}}$

   $= \dfrac{\sqrt{3}}{2 \times 3}$

   $= \dfrac{\sqrt{3}}{6}$

2. $\dfrac{5}{\sqrt{7}}$

   $\dfrac{5}{\sqrt{7}} \times \dfrac{\sqrt{7}}{\sqrt{7}}$

   $= \dfrac{5\sqrt{7}}{7}$

Teaching Points

1. Apart from the basic rule for simplifying surds, the main emphasis is to draw comparisons between surds and Algebra. A surd can be treated as an algebraic term, thus it forms its own group and is subject to the same rules as for all groups (see the Algebra section).

2. Of course, if we are going to be precise here, $\sqrt{4}$ say, is really ± 2. We ignore the negative part for simplicity. However, it may be worth pointing out this fact as it may significance in the future.

Extensions

1. The first extension relies on the fact that you have covered quadratics. It centres on the following problem:
   Rationalise the denominator of:

   $$\frac{2}{\sqrt{5} + \sqrt{3}}.$$

   The trick to getting the bottom term whole is to use the quadratic formula:
   $$(a+b) \times (a-b) = a^2 - b^2$$
   So if the bottom term is a '+' then multiply top and bottom by '-'. I.e.

   $$\frac{2}{\sqrt{5} + \sqrt{3}} \times \frac{\sqrt{5} - \sqrt{3}}{\sqrt{5} - \sqrt{3}} \quad \text{(The top line should be in ( ))}$$

   $$= \frac{2(\sqrt{5} - \sqrt{3})}{5 - 3} = \frac{2(\sqrt{5} - \sqrt{3})}{2} = \sqrt{5} - \sqrt{3}$$

2. **Further investigation of irrational numbers**
   (a) **The golden ratio or golden section**.
   If we have a line of say 1m in length and divide this line, so that the ratio of the shorter part to the longer part is the same as the ratio of the longer part to the whole length (1m). The longer part can be calculated:
   Call this longer part, x. Then the shorter part would be (1-x) and the whole length is 1.

   $$1m$$

1-x	x

   Then the golden ratio states that $(1-x)/x = x/1$
   Multiply both sides by x and we get
   $1-x = x^2$
   Or $x^2 - x + 1 = 0$
   Solving the quadratic, we get that $x = (-1+\sqrt{5})/2$ (Only take the + answer as it is a length)
   An irrational number. It is usually represented by $\phi \approx 0.618$
   This ratio was used in many areas, especially painting portraits where the ratio of the length of the portrait to its width was the Golden Ratio.

(b) **An investigation of Euler's number** e ≈ 2.718. What is it and where does it come from?

Trigonometry

Trigonometry suffers from the problem of being bogged down with way too many formulae. The approach here is to try and reduce some of those formulae.

Where does Trigonometry come from? (Some basic ideas)

Students tend to struggle to appreciate where Trigonometry comes from and what its value is. So here is one approach:

Consider a right angled triangle:

Let's say the Hypotenuse is 1m and each of the smaller sides are the same (say a for the moment)

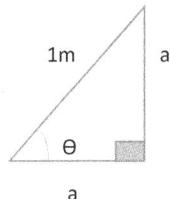

What do we know about this triangle?

1. By Pythagoras' Theorem: $a^2 + a^2 = 1$ -> $2a^2 = 1$ -> $a^2 = \frac{1}{2}$ -> $a = 1/\sqrt{2} = .7071$m
2. The two smaller angles inside the triangle must be the same because the two sides are the same. Hence each angle, Θ, must be 45°. This is because the 3 angles always add up to 180° and you already have 90° accounted for.

NOTE: In Trigonometry, we usually use the Greek letter, Θ (Pronounced Theta) for the unknown angle, rather than say x, as a tribute to Pythagoras.

Now, suppose we double all the sides, thus:

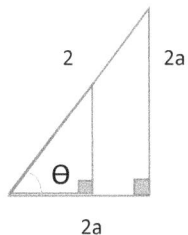

Notice that the angle, Θ, has not changed. In fact we could change the lengths of the two shorter sides, in all sorts of ways and preserve the angle

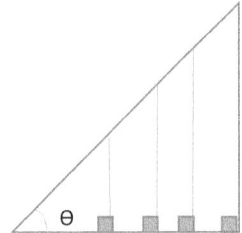

Is there a connection between the lengths of the sides and the angle?

Consider firstly the side opposite Ө and the Hypotenuse. In the first example, we doubled each of the sides. What could you do to the opposite side and hypotenuse so that when you doubled both you still had the same answer?

The easiest combination is to divide them – Opposite side / Hypotenuse. Doubling each, still gives the same answer. Similarly for other combinations (see diagram). If we reduced the Hypotenuse by say 30%, you will see that the opposite is reduce by the same in order to keep the same angle, Ө. (This is relating to the concept of similar triangles). If fact whatever we do the hypotenuse, we need to do the same to the opposite side as well. Hence the formula, opposite side / Hypotenuse will ALWAYS be the same answer.

Hence there is a direct connection between the angle, Ө, (as shown) and the ratio opposite side/ hypotenuse AND it is always the same answer. For our example the angle, Ө, was 45° and the opposite side/ hypotenuse is .7071/1 or just .7071.

We call this ratio the **SINE** of the angle. As a formula, it is written as:

$$\textbf{SIN } \Theta \textbf{ = OPP/HYP} \quad \text{(for short)}$$

NOTES:

1. The main point to note is that the Sine of an angle is a set value. For example, Sin45° = .7071. Others include Sin56° =.8290 and say Sin72°=.9511.
2. These values are fixed no matter what right angled triangle you construct due to the concept of similar triangles.
3. The values of Sine vary from 0 to 1. Can you see why?
4. The values for Sine can be found on your calculator.
5. For any angle inside a right angled triangle, we call the side opposite the angle the Opposite side (OPP) and the side next to the angle the Adjacent Side (ADJ). The Hypotenuse (HYP) is always opposite the right angle and is the longest side.

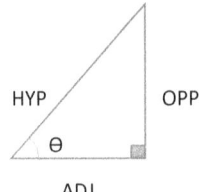

HYP   OPP

Ө

ADJ

Examples

1. Find a

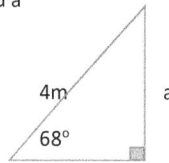

2. Find a

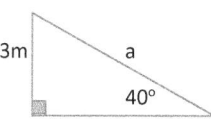

Sin a = OPP/HYP

Sin 68° = a/4

0.9272 = a/4

0.9272 x 4 = a = 3.7087m

Sin a = OPP/HYP

Sin 40° = 3/a

0.6428 = 3/a (To find a, shift it to the

left)

0.6428 x a = 3   first – Remember

nasties in

a = 3 /0.6428   Linear Equations)

a= 4.6672m

NOTE: In Example 2, did you notice how we solved the problem? A shortcut is to simply swap the position of the 'a' and '0.6428' around. For all the Sin problems, you will solve them by either using the method in example 1 or example 2.

COS & TAN

We really just picked the ratio OPP/HYP as an example. What about other ratios?

For instance does the same apply for say, the ratio adjacent side/hypotenuse?

Let's take a specific example. Using Pythagoras' Theorem, here is a typical right angled triangle

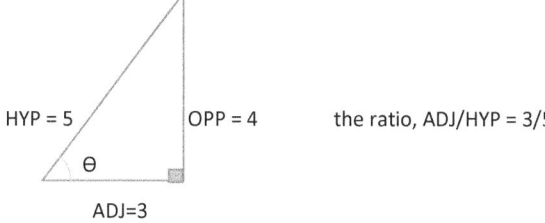

the ratio, ADJ/HYP = 3/5

If we doubled each side, we would get:

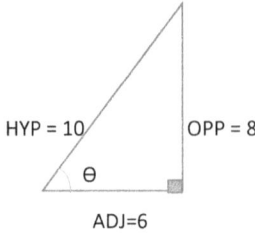

HYP = 10   OPP = 8

Θ

ADJ=6

By the theory of similar triangles, the angles inside DO NOT CHANGE. So the ratio, ADJ/HYP for this one would be - ADJ/HYP = 6/10 = 3/5.

In fact, by similar triangles we can increase or decrease the triangle by any amount and still have the same answer for ADJ/HYP/

We call this ratio the COSINE of the angle. As a formula, it is:

$$\boxed{\textbf{COS } \Theta = \textbf{ADJ/HYP}}$$ (for short)

Examples

1. Find a

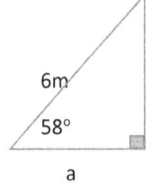

6m

58°

a

2. Find a

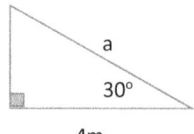

a

30°

4m

Cos a = ADJ/HYP

Cos 58° = a/6

0.5299 = a/6

0.5299 x 6 = a = 3.1795m

Cos a = ADJ/HYP

Cos 30° = 4/a

0.8660 = 4/a (To find a, shift it to the

left)

0.8660 x a = 4   first – Remember

nasties in

a = 4 /0.8660   Linear Equations)

a= 4.6188m

And yes there is a third ratio, OPP/ADJ. This is called the TANGENT of the angle. Again because of similarity, the Tangent (TAN for short) of any angle is always the same. As a formula it is:

$$\boxed{\textbf{TAN } \Theta = \textbf{OPP/ADJ}}$$

Examples

1. Find a

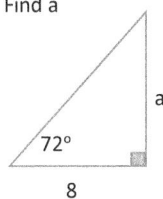

2. Find a

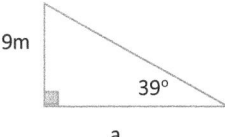

Tan a = OPP/ADJ

Tan 72° = a/8

3.0777 = a/8

3.0777 x 8 = a = 24.6215m

Tan a = OPP/ADJ

Tan 39° = 9/a

0.8098 = 9/a (To find a, shift it left

0.8098 x a = 9   first – Remember

                    nasties in

a = 9 /0.8098   Linear Equations)

a= 11.1141m

Teaching Points

1. Students can get easily confused as to which formula is which. An age old saying is to combine the first letter for each. You get

        SOHCAHTOA (pronounced "Socatoa")

Just remember that phrase and you can easily reconstruct each formula:

        SOCOHTOA = Sin Opp Hyp Cos Adj Hyp Tan Opp Adj

2. It is important that students realise that the Sin, Cos and Tan of an angle is unique. It does not matter which triangle the angle is located in.
3. A useful exercise is to look at the limits of Sin, Cos and Tan. Either by using a triangle visually or by simply looking at values, you can show that for angles 0° – 90° , the Sin goes from 0 to 1, Cos goes from 1 to 0 (i.e. it decreases as the angle increases) and Tan goes from 0 to ∞ (infinity) (Tan is the only one where answers can be greater than 1).
4. To help students remember some common Sin Cos and Tan formulae, they need only recall the following diagrams:

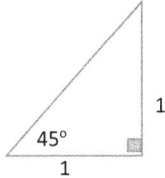

Draw a triangle with each shorter side = 1. Since both sides are the same, the 2 angles are the same, i.e. 45°. By Pythagoras, the HYP is $\sqrt{2}$.
So by the diagram,
Sin 45° =$1/\sqrt{2}$, Cos 45° = $1/\sqrt{2}$ and Tan45° = 1

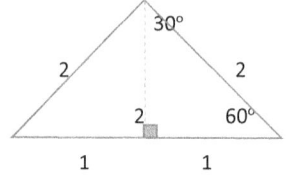

Draw an equilateral triangle with each side of 2. This means that each angle inside is 60°. If you then split it in half, you get a right angled triangle, where the third angle has to be 30° and by Pythagoras, the vertical side is √3.

From this diagram, you can calculate the following:

Sin 60° = √3/2, Cos 60° = ½ and Tan 60° = √3. And Sin 30° = ½, Cos 30° =√3/2 and Tan 30° = 1/√3.

(Hint: from the work in Surds, it is better to keep the numbers in the square root form. Also you will find that most textbooks will have questions using these specific angles)

5. The traditional unit circle approach to introducing Sin, Cos and Tan has its merits, especially if you intend to extend students into trigonometry of angles bigger than 90°. However, the similar triangle approach explained at the top has the advantage of starting from concepts you would have already covered. Namely Similar Triangles and Pythagoras.

Extension work

1. A useful exercise might be to go through the unit circle approach as an extension. It will be handy if you want to try extension exercise 4 below.
2. As I have mentioned in the explanation above, the ratios OPP/HYP etc. were simply picked as a suggestion. Actually they have been generally accepted as the main ratios to use. However, there are other combinations to consider. For instance HYP/OPP, ADJ/OPP and ADJ/OPP. These do have specific names associated with them.
HYP/OPP = Secant Θ (SEC Θ), HYP/ADJ = Cosecant Θ (COSEC Θ) and ADJ/HYP = Cotangent Θ (COT Θ). Each of these have their uses and it is an interesting exercise for students to investigate what they are.

3. Can you find a formula that connects Sin, Cos and Tan together? (Tan Θ = Sin Θ /Cos Θ). Similarly there is also a formula that connects Sin with Cos. (Sin^2 Θ + Cos^2 Θ = 1)
4. On most calculators and computers, angles are recorded in Radians as well as Degrees. Find out what a radian is and how to convert between the two units. Why use Radians? Also some calculators and computers use Gradians. What are these?

Spreadsheet Formula

No maths course is complete unless students get some instruction on how to use a spreadsheet. The spreadsheet has had a major impact on calculation, probably as much as the introduction of the calculator did back in the 80's

It doesn't matter which spreadsheet. They all perform in similar ways. The main emphasis is to show students how to create formulae, replicate (copy) formulae and how to create a chart (graph). No longer do you draw a graph by measuring out the axes and plotting each point. The spreadsheet makes it all automatic.

There are many basic introductory lessons on how to use a spreadsheet. I have included a sample set of lessons at the end of the book. The following explanations assume that you have a basic understanding of cells, formatting and editing.

Formulae (Functions)

Any cell value can be used in a formula that is typed in another cell.

Example:

If I have the value 30 in say cell B1 and I want to find out what that value times 2 plus 20 would be, I could create a formula. In say cell B3, I could type:

= 30*2 + 20 and press Enter

It should show the answer directly (80). However, it is more common to use the cell reference (i.e. the name of cell that contains the value) in the formula, rather than its actual value. So I would type instead

=B1*2 + 20 and press Enter

You should get the same answer. However, if you change the 30 into another number, say 40, the formula will auto recalculate the new answer (100)

NOTES:

1. All formulae start with an '=' sign. This tells the sheet a formula is coming after
2. In computing, times is shown as a '*', divide or fractions have a '/' and powers have a '^'.
   E.g 1. =B3/5 means get the contents of cell B3 and divide it by 5
        2. = B3^2 means get the contents of cell B3 and square it.

Spreadsheets also have a range of preset formulae called Functions. Spreadsheet Functions are written in a code format. The code is:

- A function always starts with an '=' sign
- Next is a short code that represents the formula
- Finally in brackets, we have the cell references that the function is going to be done to. For a range of cells, the first and last cells are mentioned, separated by a colon (:).

Examples:

For each of these examples, I will use the following table:

	A	B	C	D
1	Name	Test Score 1	Test Score 2	
2	Andy	47	68	
3	Bill	89	72	
4	Carrie	66	44	
5	Daniel	50	79	
6	Edgar	93	83	

1. To find the total score for Andy, go to cell D2 and type in:
   = SUM(B2,C2) and press Enter. You should get 115.

   SUM – code for finding the total; (B2,C2)- means to add up the values in cells B2 and C2

2. To find the average (mean) of all the tests for test score 1, go to cell B7 and type in :
   =AVERAGE(B2:B6) and press Enter. You should get 69.

   AVERAGE- code for finding the mean; (B2:B6)- means to find the average of ALL the values from cell B2 to B6 inclusive.

3. To find the highest score for all the test 2 scores, go to cell C8 and type in 'Highest'. The go to cell C9 and type in:
   =MAX(C2:C6) and press Enter. You should get 83 (Edgar's score)

   MAX- code for finding the maximum (largest) value in a set of numbers; (C2:C6)means include all values in the cells C2 through to C6 inclusive

4. Who passed? For a pass mark in Test Score 1, you had to have a mark of at least 55. To calculate this for Andy, go to cell E2 and type in:
   =IF(B2 >= 55, "PASS","FAIL") and press Enter. You should get "FAIL".

   The code is saying that IF the value in cell B2 is greater than (>) or Equal to (=) 55, then display the word "PASS", else display the word "FAIL"

NOTES:

1. The functions and cell references do not have to be typed in capitals. They can be entered in small letters.
2. There are many other functions. For Excel, it has a TAB called Formulas. In the Function Library section, you can open up the icons and it will display a list of functions (formulae) for various topics (Maths, Logical, Financial etc.). If you hover your mouse over anyone, it will give you an explanation and a format of that function.
3. For Google Docs, go to the Insert Menu and select Function. It's range of functions is somewhat limited.

4. For Open Office (Office Calc), click on the Function wizard (*fx* button) and locate your function on the left. Double click it and fill out the rest of it in the box on the right.
5. Which function do I use? The help guides in each provide useful information, but I generally just use Google.

Replication

Replication refers to copying the formula (function). For instance, in our example from above, you may want the SUM of each of the students test scores. You could type in the formula for each student but that would take time. Imagine if you had 120 students, not 5. Instead we replicate the formula. This is done the same way, no matter what spreadsheet you are using.

Examples

Using the table above, we are going to replicate some of the functions

1. To display the total test score for each of the students, firstly go to cell D2, where you had the original function entered. Carefully move the pointer to the bottom right corner of the cell. The pointer should change into a small plus sign, like so:

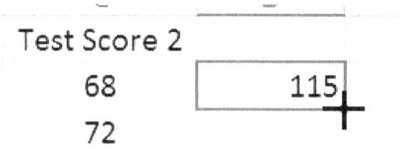

Now drag the pointer down to Cell D6, (the last score) and release. The spreadsheet should automatically enter the same function for each of the other cells. Not only that, it will also update the cell references. So check that, say cell D5, shows the formula =SUM(D5,C5). This is the powere of the spreadsheet

2. Let's replicate the average scores. Go to cell B7, where the original average formula is and again, place the pointer in the bottom right corner, so it changes to a small sign. Now drag across to cell C7 and release. The new formula will be created
3. Try it for the highest score. The original was the highest score for all the test 2 scores, located in cell C9. Move to that cell, get the 'plus sign' showing and drag back to cell B9 and release. You may also want to enter the word 'Highest' in cell B8
4. Finally, you can replicate the function for Pass and Fail. Go to cell E2 and get the 'plus sign' showing. Drag down to Cell E6 and release. Again, the function will replicate (copy) and update the cell references. To do this for Test Score 2, simply replicate the first one (in Cell E2) across to Cell F2. The replicate new function Cell F2 down to Cell F6. You now have the Pass/Fail results for both tests.

Your spreadsheet should now look like this:

	A	B	C	D	E	F
1	Name	Test Score 1	Test Score 2			
2	Andy	47	68		115 Fail	Pass
3	Bill	89	72		161 Pass	Pass
4	Carrie	66	44		110 Pass	Fail
5	Daniel	50	79		129 Fail	Pass
6	Edgar	93	83		176 Pass	Pass
7		69	69.2			
8		Highest	Highest			
9		93	83			

NOTES:

1. One of the most powerful tools of a spreadsheet is auto recalculation. You can see this by simply changing any of the test scores. All the functions will instantly recalculate the values. Try changing, say Bill's first test score to 96. After you press Enter, you should see that the highest test score has changed to 96; the average is now 70.4 and Bill's total is 168.

2. Sometimes when you replicate, you do not want the cell references to change. A classic example of this is when you are calculating percentages from a total given in a specific cell. In these cases, you precede each part of the cell reference by a $ sign. Here is an example:

    In the following sheet, we want to calculate the percentage each number is out of 120. I have put 120 in cell G1 and the list underneath:

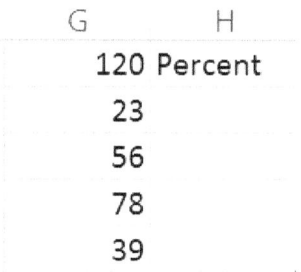

G	H
120	Percent
23	
56	
78	
39	

For the first number, 23, the function (entered in cell H2) could be

= G2/G1 * 100

Replicating the formula, you would drag down from cell H2 to H5. However, the answers are wrong. Cell H3, for instance, has a value of 243.47, which is wrong. The spreadsheet automatically updates all the cell references. So you will find that the formula, in say cell H3 reads as =G3/G2 * 100.

In the original formula, = G2/G1 * 100, we do not want the cell 'G1' to update when we replicate. To stop this from happening, we change the formula to:

= G2/$G$1 * 100

When you replicate this time, cell G1 will NOT change, giving you the right answers

Charts

Charts or graphs are a very powerful feature of spreadsheets. No more will you use pen and paper to draw up a graph.

The process is as follows:

- Create your table first complete with headings
- Highlight your table select Insert -> Chart(s)
- Select your type of Chart
- Chart will be drawn
- You can then move, resize and even delete the chart like a picture.

Example:

Using the sheet in the example from above, highlight the Name and Test Score 1 column. Include the headings. Then select Insert -> Charts. In this section, let's select a Bar chart. You should get a chart like this:

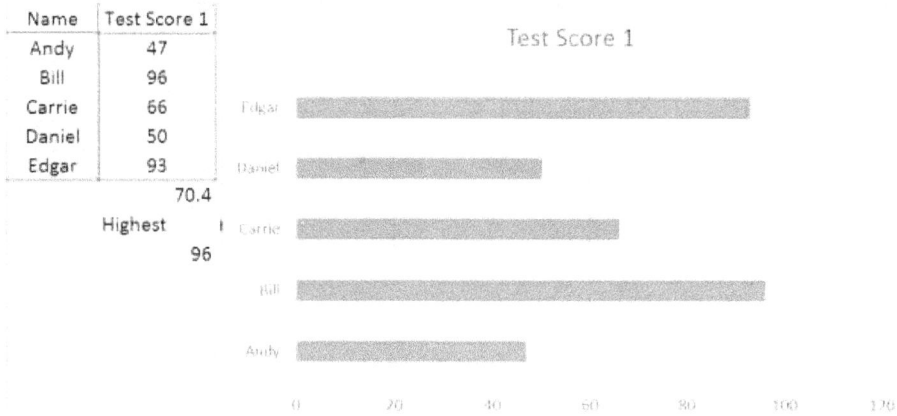

Name	Test Score 1
Andy	47
Bill	96
Carrie	66
Daniel	50
Edgar	93
	70.4
Highest	
	96

Manipulating the Chart

- When a chart is highlighted, the Chart Tools Menu is displayed.
- You can change the colour, text style, background, even the scale of the chart in here
- An alternative to this is to right click on the part of the chart you wish to change and select the Format option.

Example:

Let's change the colour of the bars to green. Add a title for each axis, and a yellow background. It will now look like this:

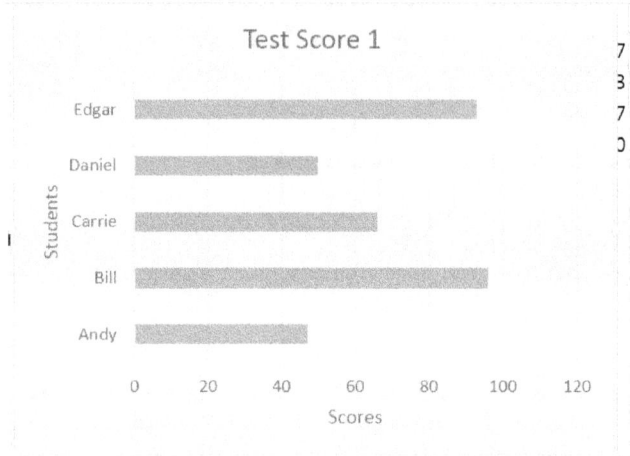

NOTES:

1. Different spreadsheets offer different options for the chart. However, for any spreadsheet, right clicking inside the chart will give you a list of options that you can use.

Line Graphs (Special Case)

Example:

Copy the following Table

Laps	Time Taken(sec)
1	56
2	67
3	51
4	49
5	55

Create a Line Graph. Have a close look at it.

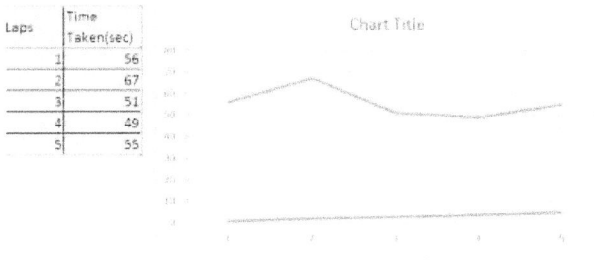

Laps	Time Taken(sec)
1	56
2	67
3	51
4	49
5	55

It will show two lines like above. This is NOT what we want. Can you see why it did this? We need to get rid the 'Laps' Line.

To do this, right click in the Chart and select 'Select Data..'.

Delete the Laps Data series. Now that's better!

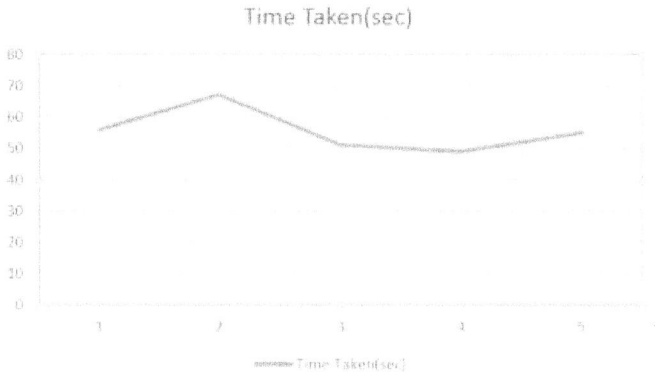

NOTES:

1. For Open Office, the option is Data Range.
2. For Google Docs, a way you can do this is to highlight the 'Laps' line, right click and select a colour of 'none'. A better way is to tick the box 'Use column A as labels', when you first create the chart.

Teaching Points

1. The spreadsheet can be likened to the calculator and used in a similar way in your Maths classes.
2. The spreadsheet has many more functions than what was shown here. For example, Look up Tables, Pivot tables and Accounting. Some of these could be investigated as extensions.

3. The main focus here is not to become an expert on the Spreadsheet but more to see how basic spreadsheet functions can make their mathematical life a lot easier.

Extension Activities

Look Up Table

This is a table that the spreadsheet uses as a reference to convert one value into another

Example:

Using our test score table from above, we might want to convert each score into a grade. Firstly insert an extra column next to each column score. It should look like this:

	Name	Test Score 1		Test Score 2	
1	Name	Test Score 1		Test Score 2	
2	Andy	47		68	
3	Bill	96		72	
4	Carrie	66		44	
5	Daniel	50		79	
6	Edgar	93		83	
7		70.4		69.2	
8		Highest		Highest	
9		96		83	

To convert each test score into a grade, we create a Look Up table. Pick another area on the sheet. For instance, the columns M and N. Let's say that for a score between 0-49 you get F, 50-59 is E, 60-69 is D, 70-79 is C, 80-89 is B and 90-100 is A. So starting at M1 and work down, we create the following lookup table:

M	N
0	F
50	E
60	D
70	C
80	B
90	A

The lookup function is then inserted next to the first test score (in Cell C2). It is:

=LOOKUP(B2,$M$1:$M$6,$N$1:$N$6) and press Enter. You should get a result of 'F'

You can then replicate down for the other test scores and repeat for Test Score 2.

NOTES:

1. The Lookup function, =LOOKUP(B2,$M$1:$M$6,$N$1:$N$6), means to lookup the value in Cell B2 (47) in the list of values from M1 to M6. Continue down the list until you reach the last number that is LESS THAN OR EQUAL to the value in cell B2. Then return the opposite value in the list N1 to N6. In this case it would have stopped at 0 and returned 'F'

2. The '$' sign in front of every cell reference is simply to make sure that when you replicate, these cells do NOT update.
3. Lookup tables ALWAYS increase down the list
4. As you can see, we did not need to include every score from 0 to 100 in column M. Only the ones that caused a change in grade.

Auto Listing

Spreadsheets include an auto listing process. An easy demonstration of this is to start creating a table of say, average monthly temperatures for the year. You would start by typing in the months of the year. This is where auto listing comes in. Enter Jan in your first cell, then Feb in the cell underneath. Rather than type in the rest, highlight both cells and use the replicate trick of putting the cursor at the bottom right of the highlight so the pointer changes to a + sign. Then drag down to reveal the other months.

This trick also works for any sequence of numbers, days of the week, years and even times of the day.

The main application of spreadsheets is to incorporate it as a tool in other Maths topics. For instance...

Statistics and Charts

Create or collect a set of raw statistics on a certain topics. Some examples include population figures for countries; lap times for racing cars; cost of housing in certain areas; temperature readings throughout the year.

From these you could calculate the average (mean), standard deviation (STDEV.P function), Median (MEDIAN function), Mode (MODE.SNGL function), the maximum (MAX function) and minimum values (MIN function). Note that for different spreadsheets, the function is spelt a little differently. The ones mentioned above are from Microsoft excel.

Graphically, spreadsheets can do scatterplots, column graphs, bar charts, pie charts, line graphs and even radar charts.

Linear equations (Goal Seek)

Yes, you can solve linear equations using a spreadsheet. Here is an example:

We want to solve   3x-2 = 34

Step 1: Select two cells, say D8 and E8. D8 represents the 'x' in the equation; E8 is the equation itself. In cell E8, enter the linear equation. Replace the 'x' by the Cell D8. So type in

=3*D8-2 and press Enter. (Don't forget to type in '*' for times)

In cell D8, you can enter a guess answer but it is not necessary.

Step 2: We use the Goal Seek function to determine the answer. For Excel, go to the Data TAB and open up the 'What if Analysis' icon. For Open Office Calc, go to Tools -> Goal Seek (for Google Docs, it is called 'Solve' but it is one of their Add-ons) . In there, select Goal Seek. Fill out the table so:

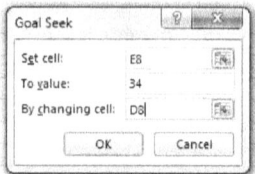

This means to set cell E8 (the equation) to the value 34 by changing the value in cell D8 (our 'x' if you like). Click OK and the spreadsheet will insert the answer into cell D8 (i.e. 12). So x= 12. The spreadsheet calculates the answer by literally trying out various numbers until it gets the right one. But it does it very quickly.

This works for just about any equation, even quadratics! Although it can only find one of the two answers and it tends to have a spurious decimal tacked on but it does work. Try it for say

$x^2 - 5x + 6 = 0$. You should get an answer of x= 3 at least.

Appendix

An Introduction to the Spreadsheet

A spreadsheet does the following:-

- Any calculation that you can do on a calculator

- Just about any type of graph

- Displays lists and tables of text and/or numbers

- Can automatically re-calculate any formula when numbers are changed

- For these exercises we will be using Microsoft Excel. However, most other spreadsheets (Open Office, Google Docs etc.) should work just as well.

Parts of the Spreadsheet:

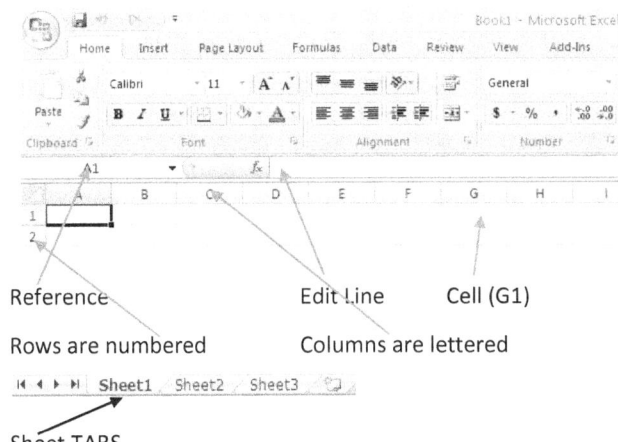

Reference   Edit Line   Cell (G1)

Rows are numbered   Columns are lettered

Sheet TABS

Each cell is identified by its Cell Reference (A column letter followed by a row number). The cursor above is currently at cell A1

Moving around the Spreadsheet:

Use arrow keys on keyboard

Click in a cell with the mouse

Press Enter after typing will move down to the next cell

Exercise:

Move the cursor around the spreadsheet, taking note of what the cell reference is. What happens if you move past the edge of the screen? How far can you go right? How far can you go down? (Hint use CTRL up arrow to return to A1)

Entering Text

- Appears on the Edit line as well as the cell
- Text is automatically LEFT aligned
- Numbers are automatically RIGHT aligned (The spreadsheet knows this is a number and thus can perform calculations on it)

Exercise: In cell F4, enter your name. In cell G6, type in your age

Editing a cell

- Simply move to that cell and re-type the correct value
- Click on the Edit Line and use it like a Text Box. Press Enter when edited.

Exercise: Copy the following table starting at cell A1:

Road Deaths over Easter

N.S.W.	12
Vic	8
Qld	6
SA	3
WA	4

Now change N.S.W. to New South Wales, Vic to Victoria etc.

Changing a cell width or Height

Place the pointer on the RIGHT HAND SIDE (width) or BOTTOM (height) of the cell. The pointer changes to a double headed arrow. Drag to the desired location. The whole column or row will change size

Exercise: Shift the State column from above so every state fits into the cell width

Deleting a cell contents

- Move to that cell. Press the Delete key

Exercise: Delete your name and age from before

Cut and paste cells

- Highlight the cells (Note: The first cell NEVER highlights)
- Use standard Cut/Paste techniques from Word

- For Drag and drop, place the pointer on the edge of the highlight (changes to a 4 head arrow) and drag.

Exercise: Move the table from above down 3 rows and across 2 columns

Insert and Delete Cells

- To Delete : Highlight the cells you want to delete.
- To Insert : Highlight the number of cells you want to insert either above or below where you want to insert.
- Right click inside the highlight and select Insert or Delete.

Exercise: Insert two complete rows between Queensland and South Australia. Add Northern Territory 2 and Tasmania 0 into these rows. Extend the column the make it fit

Formatting Cells

- Highlight the cells
- Use the Font menu as per Word

Exercise: Change the font, font size, shading and borders of the table

AutoSum – A special Formula

- Highlight the list of numbers
- Click on the AutoSum button (∑) (In the Editing Section of the Home TAB)
- A formula will be inserted underneath. It should look like this : =SUM(D3:D7). The cells in the brackets represent the range of cells to be added.

Exercise: Use the above to find out the total road deaths in Australia.

Auto Re-calculate

- Change a number and the formula automatically re-calculates

Exercise: Change the road deaths in Victoria to 10 and note what happens to the total

Final Exercise: Using another part of the sheet, create a table of the number of computers in each of five computer labs. They are Lab 1- 25; Lab 2– 21, Lab 3- 15, Lab 5- 25, Lab 6- 19. Format the table and provide a total. When done, insert the number of computers in Lab 4 -28.

Shortcuts

Formatting

- There are several preset formats for tables in the Styles Menu
- Highlight your table and select an option from the 'Format as Table Button'

- For individual Cells, Use the Cell Styles Button

Sorting:

You can easily sort columns into alphabetical and numeric order.

Exercise: Start a new sheet. Copy the following (Start at A1):

Students	Marks out of 100
Mary	56
Mark	89
Aaron	62
Betty	91
Jim	46

Highlight the table. Click on Sort and Filter (in Editing): Choose Custom filter. You will get a screen like this:

Open the Sort By Box. The headings will be displayed in there (if not click in the box marked – My Data has headers). Have a play round with the sorting options. See if you can resort the table into mark order, alphabetical order. For instance, to sort the table into mark order from top to bottom, it should look like this:

Change the student cell with the best mark into a different style using the Cell Styles Button.